THE ARCHITECT'S GUIDE TO LAW AND PRACTICE

THE ARCHITECT'S GUIDE TO LAW AND PRACTICE

BOB GREENSTREET
KAREN GREENSTREET

VNR VAN NOSTRAND REINHOLD COMPANY
NEW YORK CINCINNATI TORONTO LONDON MELBOURNE

Library of Congress Catalog Card Number: 82-24852
ISBN: 0-442-22823-6

Manufactured in the United States of America

Published by Van Nostrand Reinhold Company Inc.
135 West 50th Street
New York, New York 10020

Van Nostrand Reinhold Company Limited
Molly Millars Lane
Wokingham, Berkshire RG11 2PY, England

Van Nostrand Reinhold
480 Latrobe Street
Melbourne, Victoria 3000, Australia

Macmillan of Canada
Division of Gage Publishing Limited
164 Commander Boulevard
Agincourt, Ontario MIS 3C7, Canada

15 14 13 12 11 10 9 8 7 6 5 4 3 2 1

Library of Congress Cataloging in Publication Data

Greenstreet, Bob.
  The architect's guide to law and practice.

  Bibliography: p.
  Includes index.
  1. Architects—Legal status, laws, etc.—United
States. 2. Building—Contracts and specifications—
United States. 3. Architectural practice—United
States. I. Greenstreet, Karen. II. Title.
KF2925.G73 1983   344.73'01761721   82-24852
ISBN 0-442-22823-6   347.3041761721

Preface

In everyday practice, the architect spends considerable time carrying out various administrative tasks and dealing with problems and situations arising from the design and construction of each new building project. In order to do this completely, a basic knowledge of all the relevant procedures involved is necessary, coupled with an understanding of the broader legal and professional issues at stake.

The Architect's Guide to Law and Practice provides a comprehensive, concise, and simplified source of practical information, giving the reader a basic legal overview of the wider principles affecting the profession, and concentrating on the more specific procedural aspects of the architect's duties. In addition, it contains a series of checklists, diagrams, and completed forms which provide a quick and easy reference source.

Each section of the book culminates with a number of problems that could face the architect, laid out on ''action required'' sheets. These are dealt with in the context of a simulated office scenario on the following ''action taken'' pages where, to facilitate ease of reading, an office diary format has been adopted. The responses on these pages are not meant to be model answers, as each problem would in reality merit its own unique handling. Rather, they are meant to convey an *attitude* appropriate to successful practice management. Students preparing for architectural registration examinations are advised to work through the book, attempting the problems themselves before checking their answers against those in the text.

The 1976 edition of General Conditions of the Contract for Construction, AIA Document A201, has been referred to extensively throughout the text, and a commentary of its articles is included on pages 79 to 84. Also, many of the forms used in this book are published by the AIA; and although their use is by no means mandatory, they are useful in providing a consistency of understanding, and are therefore recommended in most cases.

The Architect's Guide to Law and Practice offers only a basic framework of information, as a detailed coverage of the numerous aspects of the subject could not possibly be crammed into 150 pages. Similarly, as most elements of law vary in each state and, in some cases, in each city, the text aims at providing a general overview of the subject. For these reasons, the text is carefully cross referenced to other sources, both at the foot of each page and in a select bibliography on page 147, which can be used for more detailed reference.

It is not the intention of the authors to provide a legal service in the publication of this book, but to offer an introduction to legal and practical matters concerning architecture. Legal assistance is advised where appropriate.

Certain texts have been referred to consistently throughout the book. These have been abbreviated as follows:

Sweet	*Legal Aspects of Architecture, Engineering and the Construction Process*
Walker	*Legal Pitfalls in Architecture, Engineering and Building Construction*
Acret	*Architects and Engineers. Their Professional Responsibilities*
AIA	*Architect's Handbook of Professional Practice*

The bibliography should be consulted for further reference.

BG/KG

Contents

THE ARCHITECT'S GUIDE TO LAW AND PRACTICE

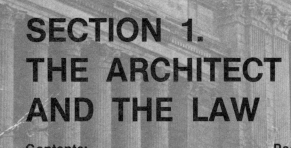

SECTION 1.
THE ARCHITECT
AND THE LAW

Contents: **Page:**

Sources of Law

The United States' judicial system developed originally from English common law, and is aimed at preserving the fabric of society. It is embodied in:

- Federal and state constitutions
- Statutes
- Common law
- Regulations of administrative agencies.

In addition, equitable doctrines are sometimes invoked.

Federal and State Constitutions

The U.S. Constitution represents the supreme law of the nation, laying down rules which bind all aspects of government. Much of its content, notably the Bill of Rights, derives from concepts which emerged through the common law.

The Constitution is the highest source of U.S. law and neither judge nor legislature may ignore or contravene its principles.

In addition, individual states have their own constitutions which are largely based upon the national model.

Statutes

Statutes are written laws officially passed by federal and state legislatures. Federal laws apply nationally, whereas state laws are only relevant to the state in which they are passed, and can vary throughout the country on the same subject (e.g., divorce).

Common Law

The basic "rules" of society have emerged through the common law which demands that judges decide each new case of the basis of past decisions of superior courts. The principle of *stare decisis* (to stand by past decisions) is not a completely rigid concept: a judge may distinguish a new case from its predecessors in certain circumstances, thereby creating a new precedent. This enables the common law to grow and adapt according to the changing values and needs of society.

Where a conflict arises between a common-law decision and a statute, the latter always prevails. Often an undesirable common-law rule is disposed of by the passing of a statute.

Regulations of Administrative Agencies

Administrative agencies are often empowered to make and enforce regulations which have the force of law.

Equity

This is a body of rules which provide a measure of fairness not always available under statute or common law. It allows additional procedures and remedies to be granted in court proceedings.

Classification of Law

Law can be classified into two basic categories:

1. Criminal law
2. Civil law.

1. Criminal Law

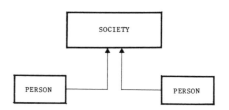

Acts committed by individuals which are proscribed by federal or state laws are generally classified as crimes (e.g., murder, theft, etc.). Lesser crimes are called misdemeanors, whereas more serious offenses are known as felonies.

2. Civil Law

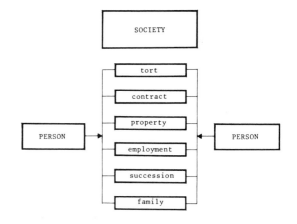

The Law

Civil law is private law dealing with the rights and obligations of individuals and corporations in their dealings with each other. Areas covered under this category include:

- Succession
- Family law
- Contract
- Property
- Tort.

The term "civil law" is also used to refer to continental European law to distinguish it from Anglo-American common law. It is important not to confuse the two usages.

For the professional design practitioner, the most relevant branches of civil law are:

a. Contract law
b. Tort.

a. Contract Law

This concerns the legally binding rights and obligations of parties who have made an agreement for a specific purpose (see page 73).

b. Tort

A tort is literally a "wrong" done by one individual (or corporation) to another for which a remedy (e.g., compensation, injunction, etc.) may be sought in the courts. Examples of specific torts are:

- Negligence (see page 7)
- Trespass (see page 54)
- Nuisance (see pages 54 and 68)
- Defamation.

It is possible for a case to fall under both contract and tort simultaneously (e.g., where a negligent act results in a breach of contract). In these circumstances, it is often easier to sue on the contract rather than attempt to prove the tort.

Reference

Sweet, pp. 1–19.

The United States has two hierarchies of courts:

1. Federal
2. State.

At the head of both hierarchies is the U.S. Supreme Court.

1. Federal Courts

Cases are heard in federal courts when a federal question is involved or when a dispute arises between parties from different states. In many cases federal jurisdiction is concurrent with state jurisdiction, but in certain matters the federal courts have exclusive authority, e.g.:

• Patent and copyright
• Actions in which the U.S. Government is a party
• Cases involving federal criminal statutes.

Federal trial courts are located throughout the United States. Each case generally begins at the District level, with the possibility of appeal to the relevant Court of Appeals and finally to the U.S. Supreme Court. Criminal and civil matters are heard in all federal courts, although certain specialized courts exist for specific issues (e.g., Court of Claims, Court of Customs and Patent Appeals).

2. State Courts

State courts are limited in jurisdiction according to their location and the type of case involved. Generally, each state has at least two levels of trial courts. Criminal matters are heard at all levels, but frequently the lowest state courts are only authorized to deal with misdemeanors.

Similarly, civil cases are heard throughout the system, but the lower courts are restricted in their jurisdiction, often on the basis of the financial amount claimed.

State court systems generally have two levels of appeals courts: intermediate courts of appeals and the State Supreme Courts. The final court of appeal is the U.S. Supreme Court.

Small Claims Courts

In many states, simple procedures have been developed for individuals wishing to sue for small amounts which would not be financially practicable in the regular court system. The financial limit for small claims varies from state to state (e.g., $1,000 in Wisconsin). In some states, professional representation is prohibited in these courts.

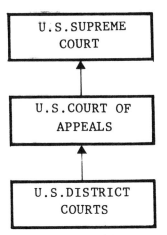

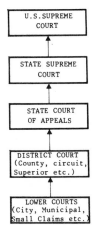

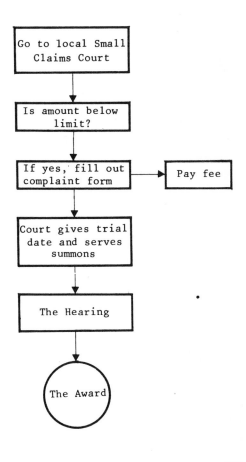

The Courts

The United States Supreme Court

The U.S. Supreme Court has original jurisdiction in cases involving disputes between states. In addition, it is the final court of appeal, but it will only hear cases it considers to be significant and which have originated in the state or federal courts.

Out-of-State Claims

Owing to federal due process requirements, some matters may be complicated if the parties are resident in different states. Many states have enacted *long-arm statutes* to enable suits to be brought against defendants resident in other states.

Standard of Proof

When a matter is decided in the courts, allegations must be proved. The standard of proof in criminal proceedings is very high: the prosecution must prove its case against the accused "beyond a reasonable doubt". In civil matters, parties need only prove their allegations to the degree that the court will accept them on a "balance of probabilities".

Other methods are available for the resolution of disputes outside the courts:

- Arbitration (see page 139)
- Administrative boards, agencies, and commissions (quasi-judicial forums which tend to be less formal than the regular courts and specialized in nature).

In most legal matters affecting design practice, it is advisable to obtain professional legal advice. Selection of an attorney may be facilitated by contacting a local or state bar association which, in many areas, operate convenient lawyer referral services free of charge.

Reference

Sweet, pp. 2–19.

The architect's legal obligations and responsibilities are owed to a variety of parties, and are governed by statutes, administrative regulations, and common law.

However, the majority of suits against architects are concerned with:

1. Breach of contract
2. Negligence.

1. Breach of Contract

Although not a party to the construction contract, the architect forms a contractual relationship with the owner (see page 43). An implied agreement is made by the architect to carry out the required work to the standards expected of the profession. Failure to meet these standards, thereby causing extra expense or delays for the owner, may result in a claim for damages against the architect on the grounds of breach of contract.

2. Negligence

Separate from any contractual obligations which may have been agreed upon, a duty of care under the law of tort may exist (see page 4). If a person fails in this duty, a negligence suit could succeed. So the architect could be liable for the consequences arising from negligent behavior even in the absence of a contractual relationship.

The extent to which any party may be held liable to others in tort depends upon their specific duty of care. In contractual situations, the obligations of both parties are usually clearly defined, but in tort it is often difficult to determine the extent or even the existence of a duty of care. However, some

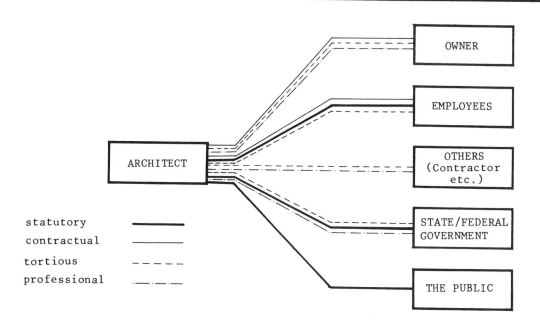

duties of care have been defined by case law and/or statute. Two of particular concern to the architect are:

- Strict liability
- Vicarious liability.

Strict Liability

In certain cases, liability may exist independently of wrongful intent or negligence. This concept is best illustrated by the English case of *Rylands v. Fletcher* (1868), in which water from a reservoir flooded a mineshaft on neighboring land and led to a successful claim for damages, although no negligence on the part of the reservoir owner was proved. The decision against the owner was made on the basis that he had kept on his land "something likely to do mischief" and that it had subsequently "escaped". This made him automatically strictly liable for the consequences.

Vicarious Liability

In some circumstances, one party is responsible for the negligent acts of another without necessarily contributing to the negligence. This is referred to as "vicarious liability" and a common example is the employer's responsibility for the acts of employees in the course of their work. A related example is the architect's liability for the defective work of consultants (see page 26).

In all cases concerning claims based on negligent behavior, certain conditions must be proved by the plaintiff if the claim is to be successful:

a. That a duty of care was owed by the defendant to the plaintiff at the time of the incident complained of
b. That there was a breach of this duty
c. That the plaintiff suffered loss or damage as a result of the breach.

The Architect's Liability

Standard of Care

In all cases, it is the "reasonable standard of care" established by common law against which a defendant's performance is matched and judged. In the case of the architect, the standard is considered to be the average standard of skill and care of those of ordinary competence in the architectural profession (see page 9).

An analysis of some specific issues in the following four pages will give an indication of the extent to which an architect may be held liable for negligent acts, and also help to highlight the areas which merit particular care and attention. It should be noted that the architect's liability in tort is subject to periodic change as a result of changes in the law and, therefore, it is necessary to be constantly aware of new developments.

Criminal Liability

In certain limited cases, individual state laws may impose criminal liability upon the architect (e.g., if death results from the violation of a compulsory building regulation which expressly states that such a situation gives rise to a charge of manslaughter; State v. Ireland 12 NJL 444, 20 A2d 69 (1941)).

References

AIA 19, pp. 5–6.
Acret, pp. 2, 44–53.
Sweet, pp. 659–660, 604–605, 678.

The Architect's Standard of Care

An architect cannot be successfully sued for negligence unless it can be proved that he/she failed to use reasonable skill and knowledge. Unfortunately, it is not easy to assess the meaning of "reasonable skill and knowledge" because this standard has emerged through decided cases involving a variety of different circumstances. Some courts have appeared to interpret the standard very strictly against the architect, whereas others have been more lenient. However, certain general observations can be made:

1. *The architect's duty as a professional person is higher than an ordinary layperson's duty.* Society expects a greater standard of care from a professional person exercising his/her particular skill.

2. *An architect is expected to exercise the care and possess the skill of those of ordinary competence in the architectural profession.* For this reason, claims against architects usually involve expert testimony by fellow architects who testify on their reactions to similar circumstances. However, courts will sometimes judge professional performance without reliance on expert testimony on the basis that no profession should, by itself, set the limits of its standard of care.

3. *An architect does not guarantee satisfactory results.* Although reasonable diligence can be expected of an architect, there is no implied warranty of satisfaction unless an express agreement is made to the contrary.

A more detailed understanding of the architect's standard of care may emerge from a careful analysis of relevant cases. However, architects should take care to maintain high standards of performance in all aspects of practice as this is the most effective defense against professional negligence claims.

Case References

Surf Realty Corp. v. Stending 195 Va 431, 78 SE2d 901 (1953)
Coombs v. Beede 89 Me 187, 36 A 104 (1896)
White v. Pallay 119 Or 97, 247 P 316 (1926)
Martin v. Bd. of Education 79 NM 636, 447 P2d 516 (1968).

Liability for Workmen

The safety of workmen on site is generally the responsibility of the contractor who, as their employer, must maintain compulsory insurance coverage under the various state Workers' Compensation Laws. These laws require that, in the event of injury to an employee, compensation is automatically paid under the employer's policy, and no negligence on the part of the employer need be proved. However, the employee is prevented from suing the employer if Workers' Compensation is received, and this has caused hardship because Workers' Compensation payments are usually lower than the damages often awarded by the courts.

This situation has led injured workmen to seek other defendants against whom a claim could be brought for damages. In the construction industry, the defendant is often the architect. In a number of recent decisions the architect has been found liable for workmen's injuries, but liability has generally depended upon the scope of the architect's authority. An architect whose site duties are merely inspection of the work to ensure compliance with the contract documents is much less likely to be found liable for a workman's injury than the architect who, in addition, supervises the method and manner of doing the work.

In some cases, architects found liable for workmen's injuries have, in turn, successfully sued the contractor in spite of the Workers' Compensation Laws.

Case References

Cutlip v Lucky Stores Inc. 22 MdAp 673, 325 A2d 432 (1974)
Parks v Atkinson 19 AzAp 111, 505 P2d 279 (1973)
Swarthout v. Beard 388 Mi 637, 202 NW2d 300 (1972).

Liability for Materials

Recent court decisions suggest that the extent to which the architect may be held responsible for negligent acts and omissions can be wide and varied. One particularly underestimated liability area is the specification of defective or unsuitable materials.

It may seem reasonable to assume that, in the specialized fields of material production and component development, the architect should not be expected to have detailed knowledge and expertise: reliance on trade literature, warranties, or certificates of compliance may seem sufficient. However, many courts have ruled otherwise. Several cases in this area suggest that the architect should be aware of the performance of products and that, in the event of subsequent failure, should be at least partially responsible for the consequences if it can be shown that there was inadequate investigation. This may have the effect of inhibiting the use of new or innovative materials in favor of traditional and predictable ones.

Pre-Specification Checklist

The following steps may be taken to provide some security against materials liability. The extent to which they are utilized will depend upon the nature of the project and the materials involved.

1. Request detailed technical information and test results from the manufacturer, together with a list of those who have already used the material in question.

2. Request details of any nationally recognized standards institutes' approvals.

3. Contact those who have used the material on a similar project to the one proposed, and ask detailed questions about its performance. If still in doubt, require independent tests.

4. Inform the manufacturer of the intended use of the material and ask for written comments upon its suitability for that proposed purpose.

5. Request warranties from the manufacturer.

6. Keep all written representations and warranties in case of future problems.

7. If installation of the material is unusual or specifically defined by the manufacturer, request that the manufacturer's representative be present when the material is installed.

8. If a specified material becomes unavailable, ensure that all necessary modifications are made to the design before installing a substitute.

Case References

Bloomsburg Mills v. Sordonic Const. Co. 401 Pa 358, 164 A2d 201 (1960)
Scott v. Potomac Ins. Co. of D.C. 217 Or 323, 341 P2d 1083 (1959).

Statutory Limitation of Liability

In all states, some form of statute of limitations has been enacted. These statutes provide a specific limited time period for bringing a legal action. Two factors are important when assessing protection under a statute of limitations:

1. The specified period, which can be discovered from the statute.
2. The time when the specified period commences. The statute may be of some help, but usually this factor develops out of decided cases.

Specified Period

The limitation period varies considerably from state to state (e.g., 3, 6, and 10 years for negligence claims), and periods are frequently longer in the case of a contractual relationship.

Commencement of Limitation Period

Depending upon the statute and relevant decided cases, the period could begin:

- At the date of substantial completion
- At the end of the professional relationship, which continues as long as the architect is dealing with the owner (e.g., site visits during the contractor's 12-month warranty period)
- At the time that the defective work occurs
- At the time the plaintiff discovers, or should have discovered the fault.

In some states, limitation periods are governed by more than one of these events: for example, in Minnesota, a civil action can be started within 2 years of the defect being discovered, but not later than 10 years after completion.

Where discovery of the fault governs the commencement of the limitation period, the architect's liability is expanded dramatically. If a plaintiff can wait until a defect appears before the specified period even begins to run, insurance in respect of future claims would have to be carried indefinitely.

The law on this matter in each state should be checked carefully by the architect, and insurance coverage maintained accordingly. Where AIA Document B141 is used, Article 11.3 provides that the period of limitation between the contracting parties begins ''not later than the relevant Date of Substantial Completion of the Work, and as to any acts or failures to act occurring after the relevant Date of Substantial Completion, not later than the date of issuance of the final Certificate for Payment''. This, of course, does not affect the rights of third parties to sue for negligence at other times in accordance with the relevant statute of limitations.

Copyright

Until 1978, copyright was governed by both common law and statute. However, a federal law which became effective on January 1, 1978, abolished common law rules of copyright, except where work was not fixed in a tangible form. To secure statutory protection of design material, the architect may follow certain procedures:

1. A provision should be inserted in the owner–architect agreement, stipulating that the drawings and specifications are instruments of service and remain the architect's property. AIA Document B141 (Standard Form of Agreement between Owner and Architect) includes an adequate provision in Article 8.1.

If the question of ownership is not made clear in the contract, a court might find that the design plans, etc., are the property of the owner. This could have serious liability consequences: for example, the owner might dispose of the design to a third party who could then sue the architect in the event of a design error.

2. A notice of copyright should appear on each item of the work in the following form:

a. The word 'Copyright', abbreviation 'Copr' or symbol ©, followed by
b. The year when the work was first published
c. The name of the copyright owner. Example: © 1982 John Green

3. In the event that the architect intends to take legal action for infringement, the copyright should be registered. All relevant documents together with the application form and registration fee should be filed at the Copyright Office, Library of Congress, Washington, DC 20559.

Ethics and the AIA

Prior to 1981, the conduct of AIA members was regulated by AIA Document J330, The Standards of Ethical Practice. This was a mandatory code which provided for disciplinary action in the event of noncompliance. The Standards of Ethical Practice have now been replaced by a statement of Ethical Principles (AIA Document 6J400), compliance with which is entirely at the discretion of individual members (see page 37).

The reason for this major change can be attributed to a dispute between the AIA and one of its members. Aram Mardirosian was suspended from the Institute for allegedly supplanting a fellow architect, and thus contravening one of the ethical standards. The supplanting rule provided that members of the AIA could not pursue architectural work upon which another architect was employed until the latter's contractual relationship with the client had terminated. In addition, the former architect had to be notified of the successor's position. However, Mardirosian refused to accept the validity of the supplanting rule which had resulted in his suspension, and he filed in Federal Court against the AIA, claiming that enforcement of the rule was a direct violation of the Sherman Antitrust Act in that it represented a restraint of trade. After a federal judge agreed with Mardirosian's argument, an out-of-court settlement was negotiated and Mardirosian accepted $700,000 and a promise that all AIA records of his suspension would be expunged.

Soon after the settlement, the AIA abolished the Standards of Ethical Practice and produced its new statement of Ethical Principles. Since the change, one of the most controversial issues has been architectural advertising, which was formerly restricted. Some firms have started to advertise their services on a large scale, but the profession is divided as to the appropriateness of such conduct.

It is important to note that, in spite of the new voluntary character of the AIA Ethical Principles, individual state laws relating to the conduct of architects are absolutely mandatory, and apply to all state-registered architects, regardless of AIA membership.

References

Sweet, pp. 842–844, 884–917.
Walker, pp. 194–203, 59.
Acret, p. 254.

The law can be seen as a complex web of rules and procedures that either control or affect the actions of individuals and groups. Breaking the rules, whether intentionally or not, might lead to the implementation of prescribed punitive or compensatory measures.

In the construction field, a number of precautions and remedies are available to prevent or allow for certain contingencies. The most important of these are:

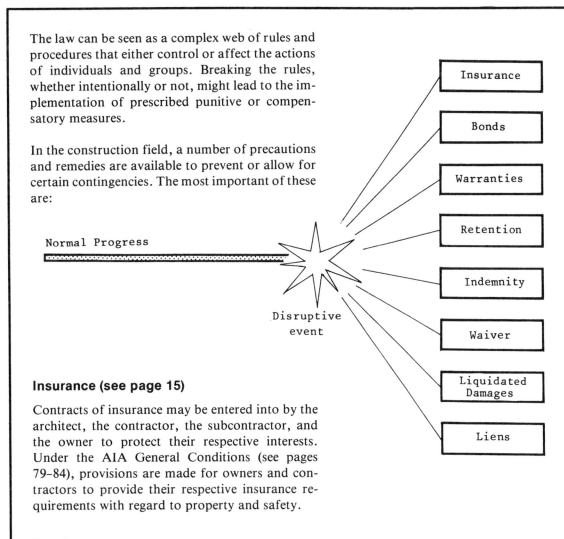

Insurance (see page 15)

Contracts of insurance may be entered into by the architect, the contractor, the subcontractor, and the owner to protect their respective interests. Under the AIA General Conditions (see pages 79–84), provisions are made for owners and contractors to provide their respective insurance requirements with regard to property and safety.

Bonds

These fulfill a similar function to insurance: they enable an owner to claim relief from the surety who underwrites the contractor in the event of the latter's noncompliance with the contract requirements. Types of bond include performance bonds, bid bonds, and labor and material payment bonds (see page 95).

Warranties

These are assurances given by parties in respect of their goods and/or services (e.g., roofing) which usually last for a stated period of time and are legally enforceable.

Retention

Before each progress payment is made during the construction phase, an agreed percentage (usually 10%) will sometimes be retained by the owner to ensure the contractor's continued performance until the completion of the work, when the accumulated sum is released. Though a prudent precaution for owners, retentions are unpopular with contractors and, in recent years, retained amounts have tended to be increasingly lower.

Variations of the procedure include retaining a percentage for the first 50% of the work only, after which the retainage, with the consent of any surety, may be reduced or discontinued (see page 95). Alternatively, an agreed percentage may be retained upon the first 50% on each line item of the work enabling subcontractors to benefit from early release. Some parties may agree to invest the retainage in order to accrue interest payable to the contractor upon successful completion of the work.

Indemnity

One party may secure or "indemnify" another against liability for loss or damage resulting from certain circumstances (e.g., AIA A201, Article 4.18). Indemnity may be implied by events but, in the construction industry, it is generally expressed in a written contract. However, legal actions against architects are frequently based on implied indemnity.

Safeguards and Remedies

Waiver

A waiver indicates the giving up by one party of rights which may prevail over others (e.g., in some instances, the acceptance of payment may constitute the waiver of certain claims against the payer). Waiver of some rights is restricted by individual state laws (e.g., waiver of lien; see below).

Liquidated Damages

These represent a formula specified by the contract documents which provides an agreed method of assessing damages arising from late completion (e.g., $x per day, to be paid by the contractor to the owner for every day by which the agreed completion date is exceeded; see page 115).

Liens

In cases where goods and/or services have been provided, the supplier may be able to secure a private *mechanic's lien* or "hold" upon the recipient's property to ensure payment of outstanding fees. The applicability of lien laws varies from state to state, particularly with regard to professional services. A lien effectively encumbers the title of the property and may be released after satisfactory settlement of the debt.

Some states allow the architect to impose a lien for design work and administering the contract, whereas other states only allow a lien for work done by the architect on site. A few states do not permit the architect any lien at all. In view of these considerable variations, individual state lien laws should be carefully noted before attempting to make use of this remedy.

Claims: Settle or Defend

If a claim is made upon the basis that legal obligations have not been fulfilled, the party so charged may admit responsibility and settle the claim by agreed damages or other appropriate means of compensation. Alternatively, the claim may be denied, in which case it is likely that the dispute will be resolved either by *litigation* (i.e., through the civil court system), or by *arbitration* (see page 139).

Shared Liability

It is possible that more than one party will be cited in a tort action on the basis that they share responsibility for the act or omission complained of. In these circumstances the cited parties may become *joint tortfeasors*.

Time Limits

Lapse of time may affect the validity of a civil court action, and individual states have promulgated limitation statutes. These vary, not only as to the time limit for bringing an action, but also as to the commencement of the limitation period (see page 11).

References

Walker, p. 59, 137–140, 204–241.
Acret, pp. 254–275, 276–291.
Sweet, pp. 428–429, 732–735, 844–845, 400–408.
AIA 17, p. 12.
AIA 10, p. 13.

A contract of insurance is created when one party undertakes to make payments for the benefit of another if specified events should occur. The conditions upon which such a payment would be made are usually described in detail in the *policy*. The *consideration* (see page 74) necessary to validate the insurance contract is called the *premium*.

Types of Insurance

The most important types of insurance relating to the construction process are:

1. Professional liability
2. Public liability
3. Construction contract.

1. Professional Liability

In the light of current statistics which indicate an alarming rise in negligence suits against the architectural profession, liability insurance is a valuable means of providing financial protection. However, there is no legal requirement to insure, and some architects prefer to risk the consequences and save the high cost of premiums. This attitude has its disadvantages because some clients may require proof of insurance as a prerequisite to employment.

Professional liability insurance (often referred to as E & O, or Errors and Omissions) varies from company to company both in coverage and conditions, and great care should be taken in policy selection. In particular, the time limitation on claims under the policy should be checked (e.g., to discover whether the policy covers errors made prior to the policy period, which only become apparent during the policy period). Joint ventures (see page 23) are not covered automatically by pro-

fessional liability policies, and at the outset of a joint venture agreement, the architect should contact the insurer to request the necessary coverage.

Even the most careful and experienced architect should consider the security afforded by professional liability insurance, particularly because:

a. Even if not negligent, the architect must still finance the defense of claims, unless protected by a suitable policy
b. The architect is vicariously liable for the errors and omissions of employees; many professional liability policies provide coverage against this.

2. Public Liability

Most architects, whether or not insured under a professional liability policy, carry a comprehensive general liability policy to protect against claims involving injury to persons or damage to property in connection with the architect's business or premises. These policies often exclude the risks specifically covered by professional liability policies. In addition, the architect in practice may require:

Employee-related insurance:
• Workers' Compensation
• Disability
• Medical
• Retirement
• Death/dismemberment
• Group life

Office-related insurance:
• Building
• Building contents
• Documents
• Business interruption
• Criminal loss
• Motor vehicles.

3. Construction Contract Insurance

In most building contracts (e.g., Article 11 of AIA A201), both parties are required to insure against contingencies relating to personal injury and property damage resulting from operations on site (see pages 79–84).

Points to Remember

Advice by the architect to the owner on matters of insurance should be avoided and may be specifically prohibited in some professional liability policies. Similarly, many types of policy become voidable if the insured fails to follow instructions prohibiting admission of liability. Policies should be read carefully to avoid potentially expensive errors.

Contracts of insurance are said to be of "the utmost good faith" (*uberrimae fidei*). This means that all material facts which might affect the insurer's willingness to accept the risk must be disclosed. Failure to disclose may render the contract voidable (see page 74).

Insurers should be notified immediately of all events which may affect the policy (e.g., changes in personnel).

Regularly check that the amounts of coverage are adequate, bearing in mind inflation, new acquisitions, etc.

Keep all policies in a safe place at the principal business location.

Ensure that renewal dates and premium payment dates are carefully noted so that policies do not lapse through inadvertence. Never take insurance

Insurance

cover for granted. If in doubt as to whether a risk is covered, check with the insurers promptly and ask for confirmation of specific coverage in writing.

Although personally unconnected with construction-related insurance policies, the architect should ensure that evidence of insurance required from the contractor has been approved by the owner prior to any certifications for payments (AIA Document A201, Article 11.1.4).

References

Acret, pp. 292–315.
Sweet, pp. 872–883.
Walker, pp. 49–52.
AIA B-1, B-2.

1. Letter with attached memo

Our Ref: TS/cc   December 8, 1980

Dear Mr. Wiley:

    re: Development of office building,
           Holdemat Bay.

It has come to our attention that final payment for service rendered in respect of your above project is still outstanding.

We would appreciate your prompt attention to this matter, and look forward to hearing from you in the near future.

                Yours sincerely,

                Fair and Square

Attached memo:

MEMO

To: Bill Fair
From: Tom Square
Date: Jan. 2, 1981
Re: Wiley's outstanding fees.

Since this letter, I sent a stronger one asking for payment within 14 days. We have heard nothing. What shall we do now?

2. Letter

Ref: AH/jr   January 2, 1981

Dear Sirs:

    re: House at Rock Bottom, Wisconsin

We are acting for Miss D. Mina in respect of her above property of which you were the architects in charge eight years ago. Last year, our client discovered signs of serious subsidence in the foundations of this house. The condition has now worsened, and considerable work will be required to restore the property to its full value.

We are at present seeking estimates in order to assess the loss which our client has suffered as a result of your negligence. Should you be unwilling to admit your full responsibility in this matter, we propose that an arbitrator be appointed in accordance with the agreement between yourselves and our client when estimates are to hand.

We look forward to receiving your comments.

                Yours faithfully,

                Howey, Blough and Stallham,
                Attorneys at Law

3. Renewal notice from Justin Case, Insurance agent

Fair and Square   January 5, 1981
Architects

<u>NOTICE</u>
Policy No. 69708437: Office Contents.

The premium in respect of this policy is due on January 16, 1981. Please ensure early payment to maintain continued coverage. If you have any questions, please contact:

                Justin Case
                Insurance Agent

Action Taken

1. Letter and diary insert

Our ref: TS/cc   January 5, 1981

Dear Mr. Wiley:

re: Development of office building,
Holdemat Bay

Further to our previous letters, we note that your final account is still outstanding to the sum of $850.00.

We regret to inform you that if payment is not received within 14 days of the date of this letter, we shall be compelled to take legal action to recover the sum.

We look forward to hearing from you,

Yours sincerely,

Fair and Square

Diary insert

date: 1/5/81
re: Wiley's outstanding fees.

Final letter sent out. Check to see if payment is made in 14 days. If not, discuss whether it is worth suing for this amount. Might consider a letter from our attorneys to shake him up, but we could handle an action in Small Claims Court ourselves. Have picked up the forms in readiness.

2. Diary insert and acknowledgment card

date: Jan. 5, 1981
re: letter from Howey, Blough and Stallham, Attorneys

Have acknowledged receipt of letter, but made no comment. Have checked out drawings, specs. and site records and they seem fine. Have informed our insurance brokers and attorneys just in case. We'll just have to wait and see what they do next—could be covered by the statute of limitations anyway.

3. Memo

MEMO

To: T.S.
From: B.F.
Date: 1/6/81
Re: Insurance renewal.

Its time to pay the premium again. I think we should increase our coverage 10% for office contents to cover both inflation and replacement value of the new word processor. Incidently, our new technician D. Taylor, will be using his car for site visits and probably isn't covered on his domestic policy—we'd better adjust our motor coverage accordingly. I'll arrange a meeting with Justin Case to sort it all out. Next Tuesday morning O.K. with you?

Acknowledgment card

date: 1/6/81

Dear Sirs:

Thank you for your letter of January 2, the contents of which we note.

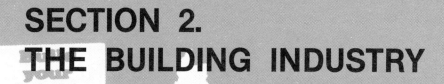

SECTION 2.
THE BUILDING INDUSTRY

Contents: **Page:**

Parties operating within the construction industry have different legal personalities according to their form of association. There are several methods of carrying on a business:

1. Sole practitioner
2. Partnership
3. Corporation
4. Joint venture
5. Other.

Before setting up any type of business, it is advisable to obtain professional legal and financial advice.

1. Sole Practitioner

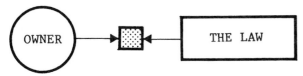

This is the simplest business form, with all liabilities and responsibilities vested in a single person. It is considered an appropriate organizational form for a small business with a predictable small-scale workload and a limited number of employees.

2. Partnership

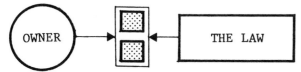

A partnership exists where two or more individuals carry on a business as co-owners for profit. All profits are shared between the partners in previously agreed proportions. The Uniform Partnership Act has been adopted by most states, and it governs the major principles of partnership law.

Partnership has become a common method of operating an architectural business as it enables architects to share their expertise, capital, and other resources.

The formation of a partnership does not limit the liability of individual partners, and each partner is responsible for all negligent acts and omissions of the firm whether personally negligent or not. However, partners joining the firm before, or leaving it after, a negligent act may be afforded protection.

Formation

The partnership relationship can be created by:

- Conduct of the parties
- Oral agreement
- Written agreement.

Most satisfactory is the written agreement, in which all aspects of the relationship can be expressed, thereby limiting the potential for disagreement or misunderstanding. In some states, all partners in architectural firms are required to be licensed architects.

Types of Partner

There are two major categories of partner:

1. The general partner
2. The limited partner.

1. The General Partner

Unless otherwise arranged in the partnership agreement, all partners are deemed to have equal rights and liabilities within the firm, and all profits of the firm are divided equally in the absence of an agreed ratio. Similarly, all authorized acts of the partners bind the partnership.

Some partnerships may agree to take junior partners into the firm. As the title suggests, junior partners have less authority and control of the business, and take correspondingly lower responsibility (usually restricted to personal acts and omissions). Profit-sharing will also be limited at this level. Care should be taken by all prospective junior partners to ensure that their position is clearly and accurately described in the written agreement. Further attention should be given to dealings with the public so as to avoid a general assumption of equality, and therefore joint liability, with the senior partners (e.g., letterheads should be clearly marked, indicating the junior partner's name and position, distinct from those of the senior partners).

2. The Limited Partner

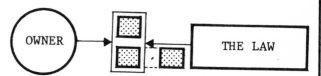

Limited partners may invest capital in a firm and share profits, but they cannot be involved in the management of the business. Unlike general partners, their liability may be restricted to the extent of their investment. Limited partners are allowed in most states under the Uniform Limited Partnership Act, but they are not common in architectural practices.

Termination of Partnership

The partnership agreement may be terminated by:

Forms of Association 1

- Expiration of an agreed time period
- Completion of a designated project or task
- Death of a partner
- Bankruptcy
- Retirement of a partner
- Mutual agreement
- Court order
- Subsequent illegality (see page 74).

Taxation

Partnerships are not taxed as distinct entities, and all partners pay individual tax upon their share of the partnership profits. Consequently, larger organizations may prefer to become incorporated in order to take advantage of tax concessions often available to corporations.

Partnership Agreement Checklist

- Date of agreement and names and signatures of the parties
- Date of termination (if any)
- Name and purpose of partnership, and business address
- Contribution of capital, provision for withdrawal, interest on capital, etc.
- Division of responsibilities and duties within the firm
- Salaries and profit-sharing details
- Methods of accounting, banking, etc., including specification of the partnership's fiscal year
- Insurance
- Benefit schemes, including pensions for outgoing partners and their families
- Rights of all partners in case of death, sickness, retirement, and withdrawal
- Arbitration agreement
- Length of vacations
- Provisions for check-writing

- Provisions for hiring and firing
- Procedure relating to loans by partners to the partnership
- Provisions in case of disqualification, bankruptcy or misconduct of a partner
- General provisions for dissolution
- Admission of new partners.

The above checklist is by no means exhaustive, and architects should note that the more detailed and specific the partnership agreement, the less chance for future problems.

References

Walker, pp. 85–94, 189–193, 258–271.
Sweet, pp. 59–63.
Acret, pp. 55–72.
AIA B-1.

3. Corporations

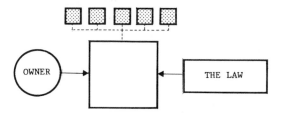

Corporations are legal entities suited mostly to larger scale operations, and owned by (although distinct from) their shareholders. Corporations can be characterized by:

- Perpetual existence independent of individual shareholders
- Profit-sharing by shareholders
- Limitation of liability of shareholders to the extent of the value of their personal share obligation (except in limited circumstances where the so-called "corporate veil" can be pierced by a court to enable an injured party to seek redress).

All corporations are subject to the law of the state in which they are incorporated. In addition, each corporation has its own Articles of Incorporation which generally draw the parameters of its activities, its organizational structure, and shareholders rights.

There are three major types of corporation:

- Profit corporations
- Nonprofit corporations (e.g., charities)
- Professional corporations.

An architect may generally be a shareholder in a corporation as long as it does not affect his professional duties. In recent years, many states have enacted statutes to enable architects to set up professional corporations in which to practice architecture.

Professional Corporations

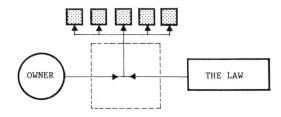

Professional corporations differ from other corporations in that, although liability can be limited in certain contractual matters, the individual professional remains personally responsible for all negligent acts or omissions despite the incorporation. Consequently, an E & O insurance policy is advisable for architects who are members of professional corporations.

In some states, the architect who practices in a professional corporation can avoid liability where the negligent act was totally outside his/her personal control. Individual state laws should be consulted to ascertain the position of members of professional corporations with regard to personal liability.

Major advantages for the architect in forming a professional corporation include certain taxation benefits, perpetual existence of the corporation, and limited security of personal assets. However, this form of association also has disadvantages such as administrative costs and formalities. Also, some public authorities may be unable to deal with professional corporations, and out-of-state work might be made difficult. For a variety of reasons, professional legal and financial advice should be sought prior to setting up a professional corporation.

4. Joint Ventures

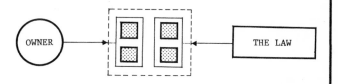

If two or more organizations wish to combine forces for a specific project, they may engage in a joint venture. This is a type of partnership limited to the duration of the task. Advantages include:

- Shared resources
- Combined expertise and knowledge
- Joint capital
- Fluidity of staff allocation.

The arrangement must be conceived as a limited one, or it may be viewed by the taxation authorities as taxable on a corporate basis. If a joint venture is felt to be an appropriate means of temporary practice, the form of agreement between the organizations concerned should be carefully drafted, specifying the precise purpose of the venture, respective tasks and responsibilities, and compensation, using the same guidelines as those for a partnership agreement (see page 22).

Formation

There are two basic types of joint venture:

- Fully integrated self-supporting joint venture
- Nonintegrated joint venture.

The fully integrated self-supporting joint venture is formed when the organizations concerned create an entirely new association, separate from the original firms, which operates independently with a separate work force, etc.

The nonintegrated joint venture is less formal and allows employees in each firm to undertake the work while remaining in their respective offices, and on the original firm's payroll. This is the more usual form of architectural joint venture.

Compensation

Firms engaged in a joint venture may divide the compensation from the venture in one of two ways:

a. Profit split
b. Compensation split.

a. Profit Split

By this method, compensation received from the owner is placed in a joint account and divided between the venturers (after expenses have been deducted) according to an agreed formula.

b. Compensation Split

This method allots a portion of the project's compensation to each venturer at the outset, and then offsets the costs of the services necessary to complete the work against the sum allotted so that the difference is retained as profit. This means that firms which operate efficiently avoid financial loss caused by the inefficiency of other firms.

In some circumstances, architects will form joint ventures with a view to being commissioned for a particular project. Rather than undergo the full requirements before the work is assured, the details of the proposed venture may be written down in a *memorandum of understanding*. This memorandum could form the basis of a full joint venture agreement if the firms are granted the commission.

Insurance can be taken out under each firm's existing policies with an appropriate endorsement, or by a separate policy in the name of the joint venture.

5. Other Associations

Other forms of organization which may be encountered in the construction industry include:

a. Associated architects, or "loose groups"
b. Professional associations and unincorporated associations
c. Trade unions
d. Governmental agencies (federal and state).

a. Associated Architects

The term "associated" with regard to architectural practice is vague, and may refer, among other things, to independent organizations sharing facilities, or to a nonintegrated joint venture of firms. The AIA recommends that the use of the term "associated" should be avoided unless the actual legal relationship of the parties is clearly defined. In the absence of a clearly defined relationship, a partnership may be implied by the courts, leading to complex and expensive liability problems.

b. Professional Associations and Unincorporated Associations

The professional association is not technically a corporation, but is sufficiently corporate to be treated as such for taxation purposes. Unincorporated associations (e.g., social clubs) are not legal entities as such, but in most states they do have limited legal capacity (e.g., to contract). Architects working for such groups should be careful to check the authority and liability of the members they deal with; this information can usually be found in the constitution or regulations of the association. State laws regarding the legal capacity of these associations should also be checked by the architect before entering into a contractual relationship.

c. Trade Unions

These are groups formed within the trades (often as unincorporated associations) for the purpose of collectively bargaining for pay and conditions of employment.

d. Governmental Agencies

The regulation of these bodies, both at state and federal level, derives from statutes. They have, in the past, enjoyed immunity from legal actions. However, this immunity is now less absolute in many states, and some claims have been made successfully against governmental agencies for their negligent acts or omissions (e.g., negligent plan inspection).

References

Sweet, pp. 63–71, 664–666, 735–738.
Walker, pp. 189–193, 272–293.
Acret, pp. 223–229.
AIA B-1, 10.

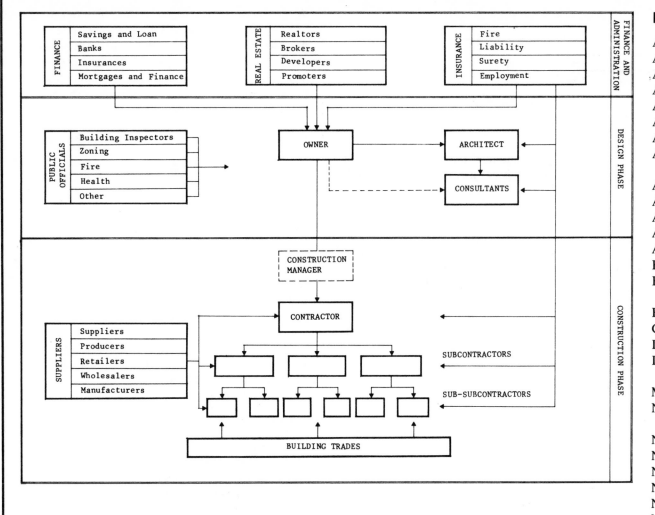

Related Organizations

American Arbitration Association
American Institute of Interior Designers
American Institute of Landscape Architects
American Insurance Association
American National Standards Institute, Inc.
American Society for Testing and Materials
American Society of Civil Engineers
American Society of Heating, Refrigerating and Air Conditioning Engineers, Inc.
American Society of Landscape Architects
American Society of Planning Officials
American Society of Mechanical Engineers, Inc.
Associated Builders and Contractors, Inc.
Associated General Contractors of America
Building Materials Research Institute
Building Officials & Code Administrators International, Inc.
Building Research Institute
Chamber of Commerce of the US
International Conference of Building Officials
Interprofessional Commission on Environmental Design
Mortgage Bankers Association of America
National Association of Architectural Metal Manufacturers
National Association of Building Manufacturers
National Association of Home Builders
National Building Products Association
National Bureau of Standards
National Constructors Association
National Fire Protection Association
National Housing Producers Association
National Society of Interior Designers
National Society of Professional Engineers
Society of Architectural Historians
Southern Building Code Congress

Architectural Relationships

The Architect/Owner

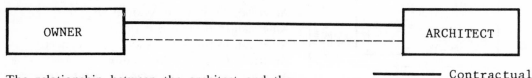

The relationship between the architect and the owner is primarily contractual, and as such is governed by the terms of the contract between them. The contract formalizes a relationship of *agency* in which the architect (the agent) acts as the representative of the owner (the principal), working solely in the latter's best interests.

Agents are expected to work with the level of skill normally associated with their profession or occupation, and to be concerned to prevent any conflict arising between their own interests and those of their principal. The agency authority of the architect is limited by the terms of the appointment, and the architect should be careful to avoid overstepping his/her authority. For example, ordering the contractor to undertake work where the latter acts upon the apparent rather than actual authority of the architect may constitute a breach of the architect/owner agreement. Should the owner wish to extend the powers of the architect (e.g., AIA Document A201, Article 3.3.1) to enable the undertaking of specific tasks otherwise outside the scope of authority, written authorization should be obtained by the architect before carrying out such work.

The agency relationship between the owner and the architect is not a general one, and the architect may act as the owner's representative only in the areas specifically stated in the contract between them. Where a decision is needed on a question in which the agent does not have authority, the principal should be contacted. In an emergency, where the

Contractual

--------- Tortious

principal is not available, the agent is authorized to do anything which prevents loss to the principal. Such situations may give rise to dispute, and should be treated with the utmost caution.

Under the AIA Contract for Construction, the architect takes on a secondary role of quasi-arbitrator of the agreement between the owner and the contractor. Absolute fairness should be exercised in this role and, in spite of being the owner's agent, the architect must not show undue favor to the owner in the event of a dispute concerning the contract (see page 107).

The Architect/Consultant

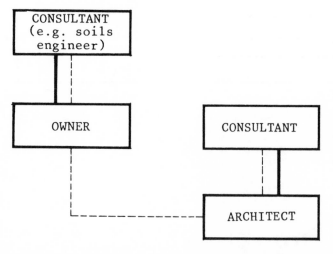

Where services necessary to a construction project are outside the architect's basic services (see page 46), specialists may be employed by either the architect or the owner to undertake the work. It is usual for the architect to form a contractual relationship with a consultant although, in some instances, it may be possible for the owner to contract directly with the specialist (e.g., soils engineer).

Types of Consultant

Consultants may be employed:

- For their technical knowledge (e.g., lighting, acoustics, landscaping)
- For their knowledge of specific building types (e.g., hospitals, theaters, schools)
- For other attributes relevant to a specific project (e.g., financial expertise, behavioral studies).

Care should be taken when employing consultants not to utilize their services for work which may fall under the architect's basic services, as this may result in reduction of the architect's fees.

Selection

As the architect is vicariously responsible for the errors and omissions of the consultants, selection should be made with great care. Owner's recommendations may be considered, but the final choice should remain with the architect, who can and should require all consultants to maintain E & O insurance policies.

In order to fully delineate responsibilities, duties, and conditions of the relationship between the architect and the consultant, a written contract is advisable. The AIA produces two standard forms which are recommended:

- AIA Document C141, Standard Form of Agreement between Architect and Engineer (see page 28)
- AIA Document C431, Standard Form of Agreement between Architect and Consultant for other than Normal Engineering Services.

These documents are written to correspond with other AIA contracts (e.g., B141, A201, etc.) in terms of timing, format, and sequence. If a consultant's services are employed, the architect may be entitled to further payment to cover administration and extra risk. In some cases, the extent of work to be undertaken by a consultant may make it appropriate for the parties to engage in a joint venture (see page 23).

For limited or clearly defined work, a carefully drafted letter may serve instead of the full contractual documents. The letter should be sent to the consultant in duplicate with instructions to return one copy signed to the architect, and it should include:

- The names of the parties
- Date of the agreement
- Title and location of the project
- Description of the work
- Terms and conditions of service
- Payment type, method, and amount.

The Architect/Contractor

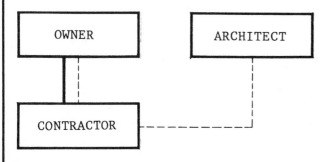

There is no contractual relationship between the architect and the contractor, as the latter contracts directly with the owner. However, most building contracts contain provisions enabling the architect to undertake prescribed duties in the capacity of the owner's agent (see pages 79–84).

Errors made by the architect which cause loss to the contractor could not result in an action under contract law (see page 73), but could form the basis for a claim against the owner who remains responsible for the agent's authorized acts. This may in turn lead to an action by the owner against the architect for breach of the contract between them. Alternatively, the contractor could sue the architect in tort, where no contractual relationship is necessary (see page 4).

The same situation arises between the architect and subcontractors whose contracts are with the contractor, and also the suppliers who deal directly with the contractor and subcontractors.

References

Sweet, pp. 45–58.
Walker, pp. 53–57, 82–83.
Acret, pp. 116–122, 125–126.
AIA 10, 11.

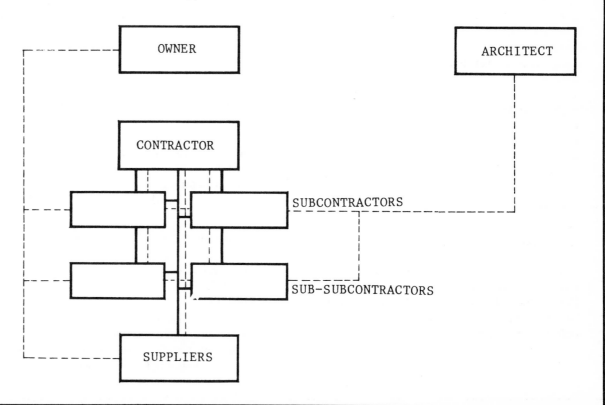

The Engineer and Construction Manager

The Engineer

As in the profession of architecture, engineering work and the title "engineer" are usually protected under state law, although often the boundary between architecture and engineering work is ill-defined. In some states, engineers may be allowed to undertake work which might be considered to be architectural elsewhere, in addition to work primarily classified as engineering.

In any event, the professional engineer will normally be expected to conform to the examination, registration, and professional requirements of the state of residence, and will be subject to many of the practice-associated conditions which apply to architects. The term "engineer" is a general description of many distinct fields of expertise, several of which are represented by their own professional bodies (e.g., the American Society of Civil Engineering). Engineering fields include:

- Soils
- Structural
- Mechanical
- Electrical
- Acoustic
- Highways
- Civil
- Drainage.

Architects and Engineers

Where architectural firms wish to engage the services of an engineer, it is advisable to use AIA Document C141, Standard Form of Agreement between Architect and Engineer. It is important to define the engineer's services as fully as possible in the contractual agreement, so that relative duties and liabilities can be determined and insurance

coverage maintained accordingly. This is particularly relevant because, although the engineer must perform to the standard expected of his/her profession, the architect is usually vicariously responsible for the engineer's negligent acts or omissions.

The Construction Manager

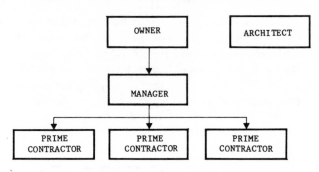

Using the services of a construction manager in the building process is a fairly recent phenomenon and, as yet, has required no generally accepted definition. The scope and detail of operations carried out under this title are varied, but tend to include:

- Design review for construction feasibility
- Cost control
- Time scheduling
- Construction methods.

It is, perhaps, for the role of construction work organizer that the construction manager has become most widely known. Where expertise in time scheduling or sequencing of operations is considered necessary, the services of an on-site supervisor acting (usually) as the owner's agent has proved useful. The construction manager has been particularly beneficial in the separate contract system (see page 85), by enabling the services of the (see page 85) various contractors to be coordinated to ensure a smooth transfer from one operation to another.

Organization

Construction management is not practiced in any conventional way; some general contracting companies have entered the field, either in addition to or instead of normal construction work. Also, architects, engineers, and others with expertise and experience in the construction industry (e.g., construction superintendents) have undertaken similar services. The contractual arrangements made with a construction manager may also vary. Often, the contract is made directly with the owner, and the construction manager acts as a go-between for all the parties involved in the building project and the owner. However, it is possible for such a manager to be employed as a consultant by the architect, or to form a joint venture with the architect (see page 23). (see page 23)

Responsibility

Despite the flexible nature of construction management at present, it has become a valuable service to the building industry. One of its major functions concerns the coordination of work and time sequencing. This carries with it a correspondingly high level of liability for actions related to supervision, and so architects involved in construction management face a higher risk of malpractice suits. Architects who offer services in this area (e.g., as additional services) should be careful to ensure that the scope of work and attached responsibilities are adequately defined in the contractual agreement, and that insurance coverage is correspondingly broad.

Standard AIA Forms that have been developed for use in these circumstances include:

- A101/CM, Owner/Contractor Agreement Form, Stipulated Sum, Construction Management Edition
- A201/CM, General Conditions for the Contract of Construction, Construction Management Edition
- A311/CM, Performance Bond and Labor and Material Payment Bond, Construction Management Edition
- B141/CM, Standard Form of Agreement between Owner and Architect, Construction Management Edition
- B801, Standard Form of Agreement between Owner and Construction Manager
- G701/CM, Change Order, Construction Management Edition.

References

Sweet, p. 103, 811.
Walker, pp. 169–174.
AIA 10.

Action Required

2. Letter

February 12, 1981

Dear Sirs:

Following our meeting last week, I would like to have the grounds of my new place landscaped professionally as you suggested. Can you do the work, or will I have to employ a specialist? I look forward to your reply,

Yours

B. Careful

1. Memo

MEMO

To: Bill Fair
From: Tom Square
Date: Feb. 11, 1981
Re: Joint venture agreement

I met with the senior partner of Morsteel & Morsteel, structural engineers, yesterday regarding the proposed joint venture with us for the upcoming school project (if we get it). I suggested that we retain the staff on the respective payrolls of our two firms and split the work between us. Shall be put together an agreement?

MEMO

To: Tom
From: Bill
Date: 2/12/81
Re: Dee Zeiner, new partner

Dee informs me that she is keen on our proposal to take her into the partnership as a junior. Before I get the new agreement drafted, what do you think are the main points to be covered—Dee is particularly concerned about her position regarding her personal liability.

3. Memo

2. Letter and memo

February 13, 1981

Our ref: BF/cc

Dear Mr. Careful:

Thank you for your letter of February 12. Work connected with landscaping falls outside our basic services as outlined in your copy of the Owner/ Architect Agreement, and comes under the category of additional services. We could certainly undertake the work for you but, as time is an important factor to you in the completion of the project, we would suggest that a firm of landscape architects be employed instead.

With your approval, we can arrange to employ such a firm and coordinate their services for you. Unless you have someone in mind, could we suggest the services of Tree Inc.? We have dealt with them in the past, and they have a good reputation in the area.

We await your instructions and will act promptly upon your written reply.

Yours sincerely,

Fair and Square

Memo

MEMO

To: Tom
From: Bill
Date: 2/16/81
Re: B. Careful project/landscaping

This letter was sent out yesterday. Careful just phoned to say go ahead. I've asked for written confirmation and have the Architect/Consultant Agreement ready. Tree Inc. want the work, so can you fill out the forms and send them off?

Diary insert

Date: 2/16/81
Re: Dee's partnership

Call insurance brokers and tell them about the new arrangement. When details of our new partnership agreement are drafted, transfer a copy to the Personnel Policy Manual for reference in the event of further junior partners.

1. Memo

MEMO

To: Tom Square
From: Bill Fair
Date: 2/13/81
Re: Joint venture with Morsteel & Morsteel

The nonintegrated joint venture sounds best, and I have dug out a list of things to be included in the agreement. Why not wait until we find out if we have the contract before having it drafted, though? A simple memorandum of understanding between the two firms will suffice for the moment. Then, if we don't get the job, no work will have been wasted.

To: Bill
From: Tom
Date: 2/16/81
Re: Dee's partnership

Got your note about Dee. Suggest the following points should be incorporated into the partnership agreement to clarify our position regarding junior partners:

Details of profit sharing
Details of any contributions to the partnership
Careful definition of respective responsibilities and duties (remember insurances)
Requirement of continued membership to the AIA and state registration
Clear definition of the junior partner status on our letterhead, brochure, and office signs.

3. Memo and diary insert

SECTION 3.
THE ARCHITECT
IN PRACTICE

Contents:	Page:

Each state enforces licensing laws which control the use of the title "architect" and the practice of architecture. Qualification requirements differ depending upon the standards of the state licensing boards, although the National Council of Architectural Registration Boards, with support from the AIA, encourages uniformity in educational and examination procedures. A combination of higher education and architectural experience is generally required as a prerequisite to a license, together with other specified requirements. The state laws are enforced by regulatory boards (see page 38).

Interstate Licensing

As the licensing of architects is carried out on a state-by-state basis, it may be necessary for the architect to requalify a number of times if engaged in work in several states. However, some states operate a reciprocity system to facilitate interstate practice, and enable architects to obtain a temporary or permanent license. Partnerships (see page 21) should ensure that all general partners are registered in states where work is undertaken. Failure to abide by state licensing laws renders the offender liable to imprisonment or fine, and may provide sufficient grounds for the owner to successfully avoid payment of fees.

The American Institute of Architects

The AIA was founded in 1857 and is the national organization representing licensed architects who are U.S. residents and who wish to be members: membership is not obligatory, but is recommended. The Institute seeks to uphold the standards of the profession, to help in the training of new architects, and to act as a corporate voice for the benefit of its members and the advancement of

"good design." The national headquarters of the AIA is located in Washington, D.C., and regional offices are maintained throughout the United States. In addition, there are state societies and local chapters. Members are designated by the letters "AIA" which may be used after their names. In certain circumstances relating to personal achievement, a member may be made a Fellow of the Institute (FAIA).

Structure

The AIA National Headquarters is headed by a Board of Directors which is composed of the Institute's officers, regionally elected directors, and the President of Student Chapters. The Board decides questions of policy and budgetary matters, while the executive committee, composed of the Institute's officers, deals with matters allocated to it by the AIA bylaws, or by the Board of Directors. The National Headquarters is divided into 8 departments:

1. Education and Research

With its dual role, this department collects and distributes relevant information to AIA members, and it also liaises with high schools, members of the profession and private and governmental agencies in connection with initial and continuing architectural education.

2. Institute Services

This department helps to coordinate the AIA internally to ensure communication at all levels. It is a highly administrative department and, among other things, it deals with membership applications and professional conduct issues.

3. Public Relations

This department attempts to induce awareness and receptivity within the public toward the profession through advertising and public relations activities. It also produces the AIA newsletter, "MEMO," which is designed to keep AIA members informed of Institute concerns.

4. Business Management

This deals primarily with financial and budgetary matters.

5. Government Affairs

In order to implement the Institute's public policy nationally, members of this department liaise with federal agencies and Congress, lobby on behalf of the profession, and provide information to architects regarding federal agency programs, regional development, etc.

6. Community Services

This department is concerned with the profession in relation to its responsibility to society at large. It has developed programs in community development (e.g., AIA/VISTA) and Equal Opportunity Education (e.g., the AIA Minority Architects Scholarship Program, Human Resources Council, etc.).

7. Professional Services

This department's responsibilities involve matters relating to architectural practice, the production of AIA Contracts and Documents (see page 105), technical services, international relations, historic preservation, and information concerning codes and standards.

8. Publishing

The department's publications include the *AIA Journal,* AIA Documents, and also books of architectural interest.

Other activities of the AIA include:

The AIA Foundation

This exists to deal with donations, awards, and prizes, and to promulgate useful information to the profession. It is a nonprofit corporation.

The College of Fellows

This comprises all the Fellows of the Institute and, among other things, the College sponsors publications and functions as an international architectural relations society.

The National Committees

These provide raw material for AIA projects in the form of committees, juries, or task forces, and they also ensure effective communication between the Institute and its members throughout the United States.

References

Sweet, pp. 793–827.
Walker, pp. 62–69.
Acret, pp. 45–46.
AIA 3, 4.

In order to maintain high levels of professionalism in both the performance and behavior of architects, certain standards have been developed to act both as guidelines and restraints. Canons of ethical behavior have been produced by the AIA to apply to all its members, and also individual state laws control the activities of architects within their jurisdiction. Architects are expected to aspire to high ethical standards in the interests of the profession and the public at large.

AIA Ethical Principles

Until March 1981, the AIA controlled the actions of its members through AIA Document J330, Standards of Ethical Practice. However, since the Mardirosian case (see page 11), the mandatory standards have been replaced by Ethical Principles (AIA Document 6J400) which are solely guidelines for voluntary adoption by AIA members. There are 12 basic ethical principles:

1. *Members should accept the primacy of learned and uncompromised professional judgment over any other motivation in the pursuit of the art and science of architecture.* This basically defines the concept of professionalism.

2. *Members should conform to the spirit and the letter of all laws governing their professional affairs.* This stresses the importance of conformity with licensing laws.

3. *Members should uphold the credibility and dignity of the profession.* Excessive advertising or self-endorsement is discouraged.

4. *Members should thoughtfully consider the social and environmental impact of their work.* Public awareness, environmental quality, conservation, and preservation are encouraged.

5. *Members in all their professional endeavors should support human rights and should not discriminate against others.*

6. *Members should be candid and truthful in their professional communications.* Truth in advertising and communications is encouraged to avoid misleading clients as to the architect's capability. Furthermore, competition sponsors should be informed of the recognized professional standards that should apply to all competitions.

7. *Members should serve their clients or employers in a thorough and competent manner.* Architects should only undertake work which they can perform competently.

8. *Members should respect the confidences of their clients, employees, and employers.* Clients should be able to rely on confidentiality in their dealings with architects, although this is not a legal requirement and may be waived in certain circumstances.

9. *Members should disclose to a client or employer any circumstance that could be construed as a conflict of interest and should ensure that such conflict does not compromise the interests of the client or employer.* Conflicts of interest may lead to unprofessional conduct: the serving of two masters whose interests conflict is discouraged. Any conflicts of this nature should be avoided or disclosed.

10. *Members should acknowledge, respect, and give appropriate credit for professional contributions of employees, associates, and colleagues.* Due credit should be given to those who contribute work to projects which are undertaken as a team effort.

11. *Members should compete fairly with other professionals and should not offer or accept any bribe or improper contribution or gift to obtain or grant work or to influence the judgment of others.* Competition between architects should be fair and open, based upon professional merit. Any activities involving bribery or undue influence are considered to be "professionally unacceptable." Agents and employees of architects should be guided by the same ethical principles, and architects are responsible for their conduct.

12. *Members should maintain and advance their knowledge of the art and science of architecture, respect the body of past accomplishments, and contribute to its continued growth.* All aspects of professional practice should be actively advanced by practitioners who should seek excellence in research, training, and practice. Furthermore, students of architecture should be given as much help as possible in their training.

The above is merely an outline of the AIA Ethical Principles, and AIA members should take care to familiarize themselves with AIA Document 6J400 which can be obtained direct from the Institute.

State Requirements

The AIA Ethical Principles apply only to members of the Institute, but every registered architect must abide by the statutory requirements of the state in which he/she is licensed. These regulations vary from state to state, but tend to reflect the concerns expressed by the NCARB in its recommended

Professional Ethics

Rules of Conduct. Individual state regulations may address the following issues:

- Lack of ability or fitness (e.g., a criminal record, insanity, violation of laws and regulations)
- Not keeping up to date with new developments and resulting lack of competence
- Conflicts of interest
- Misleading advertisements or publications
- Prevention of "fee touting" and exerting unacceptable influence on existing or prospective clients
- Duty to inform the authorities about any known unauthorized practices
- Duty to inform the authorities about any nonarchitects holding themselves out as qualified
- Prevention of stamping plans other than one's own.

Such matters are usually drafted into statutory codes (e.g., the Wisconsin Administrative Code), and they are enforced by a Regulation Board which is delegated sufficient power to impose penalties for noncompliance. Penalties include:

- Imprisonment (this is very unusual)
- Fine
- Revocation of license (either temporarily or permanently)
- Reprimand.

Reference

AIA B-8, B-9.

Although architectural offices vary considerably in their structure, management, and workload, certain general observations and recommendations can be made regarding their administration.

Administration

Initial factors to be considered in the running of an architectural office include:

1. Insurance
2. Financial management
3. Office organization
4. Staff organization and selection.

1. Insurance

As well as sufficient insurance to cover negligent performance by all employees for whom the principals remain vicariously responsible, insurance policies should be maintained to cover the office and its contents in respect of loss or damage, and also employee benefits (e.g., medical expenses, motor insurance, and compulsory coverage under the Workers' Compensation Laws). Employers will also be responsible for office safety under state and federal law (e.g., Occupational Safety and Health Act).

2. Financial Management

The necessity for maintaining accurate accounts cannot be overly stressed, and professional assistance may be necessary to establish the accounting system to be used and, possibly, to operate it. The AIA has prepared a manual for its members on the subject entitled *Financial Management for Architectural Firms* which architects in private practice may find of interest. The manual offers procedures and advice, including the use of AIA Documents (F and G series) to record activities and provide comprehensive and coordinated office records (see page 105). The financial accounting system may be organized in several ways:

- The Minimum System which is a partial accounts system.
- The Basic System which provides a level of accounting recommended by the AIA as the minimum for architects.
- The Complete System which is the most efficient and comprehensive method and is recommended by the AIA to all its members.

3. Office Organization

The physical design of architectural offices is important both in terms of the efficient internal running of the practice, and the impression given to clients and business associates who may consider it to be a reflection of the architect's design ability.

To ensure efficiency and consistency within the office's operations, it is advisable to record and maintain uniform procedures and techniques of office management. For example:

- Standardized communication methods (see page 41)
- Standardized drawing conventions (see page 57)
- Explicit roles, duties, and responsibilities for all personnel
- Use of standardized forms for office administration (see page 105). The AIA produces several of these including:
  G801 Application for Employment
  G804 Register of Bid Documents
  G805 List of Subcontractors
  G807 Project Directory
  G809 Project Data
  G810 Transmittal Letter.

Other standardized paperwork may be used including accounting forms, telephone message and memo pads, and order forms which can be printed in the house style.

In order to communicate and record information regarding office procedures in a consistent and readily available format, firms sometimes produce an "Office Standards Manual" containing the above data to provide a useful reference to new employees.

4. Staff Organization and Selection

Many practices consider it useful to establish office policy concerning their employees. A "Personnel Policy Manual" is advised as a method of consolidating preferred practice both for existing members and prospective employees to familiarize them with office characteristics and expectations. The Manual may contain general details of the practice (workload, direction, etc.), its organization, and fundamental policies regarding employment procedures and staff benefits. Information may include:

Office Practice:
- Office hours
- Payment methods
- Overtime and time recording
- Lunch and coffee breaks
- Travel and expenses
- Responsibilities (equipment, etc.)
- Salaries and salary review
- Other concessions (parking space, etc.).

Staff Benefits:
- Pensions
- Profit sharing and trust funds
- Holidays

The Office

- Incentive pay
- Sick leave
- Professional activities (further training, conferences, conventions, etc.)
- Civil duties (jury service, voting, civil projects, etc.)
- Dues to professional and civil organizations
- Future promotional policy.

Hiring practices may also be included in the Manual:

- Methods of selection
- Moving expenses and transfers
- Termination of duty, layoffs, and resignations
- Leaves of absence (military, emergency, etc.).

Contracts of Employment

A contract of employment need not be formulated in writing, but a carefully drafted agreement, covering all relevant issues established in the Office Manual will help to lessen the risk of future misunderstanding. Such agreements may be drafted into a letter of appointment, or attached to a letter in a standardized format, including the following details:

- Names of parties
- Date upon which employment commences
- Salary and payment intervals
- Hours of work
- Holiday period and holiday pay
- Sickness pay
- Pensions and other benefits
- Insurance coverage (professional indemnity, health, accident, etc.)
- Periods of notice
- Job title, duties, and responsibilities

- Required membership in professional associations
- Office benefits (e.g., automobiles).

Finally, an architectural practice should be kept under continual review with regard to procedures, personnel, and equipment. In this way, timely adjustments can be made to assure the smooth running of the firm in the event of changed circumstances.

Reference

AIA B-1.

Regardless of the type of architectural practice, efficient communication is an important factor as it can promote:

- Internal efficiency
- An external image of competency and professionalism.

Internal Communication

Many sophisticated techniques and types of equipment have been developed for the purpose of creating efficient communication. Architects must decide (perhaps with professional financial assistance) how cost-effective such technology will be in terms of their individual practice.

Whatever the size and workload of the practice, there are a number of economical procedures that may be efficiently employed:

- Date stamp all incoming mail.
- File incoming mail together with a copy of replies in chronological order.
- Develop a file of standard letters to be used in specified circumstances. This saves time and provides a consistent office style.
- Portable dictating machines can be invaluable timesavers and memory aids. They allow detailed notes to be recorded while still fresh in the mind (e.g., on site).
- Window-type envelopes can save secretarial time by obviating the need to address letters twice.
- Pre-printed acknowledgment cards can be economical but should be used with care to avoid giving an impersonal impression.
- Pre-printed internal memo pads, telephone message, and interview records are useful in helping to create a consistent recording system and conscientious record-keeping.

- Written memoranda of all important oral communications should be kept on file where possible.

External Image

The architectural firm's public image is important, as it can promote prestige among the professional community, give confidence to existing clients, and attract new ones.

The firm's image will be conveyed by:

1. Personal contact
2. Promotional literature (e.g., brochures)
3. Public involvement (e.g., lectures).

1. Personal Contact

An organization is frequently judged by its representatives, where contact will be both:

a. Oral
b. Written.

a. Oral Communication

A good interview technique is important, and the architect should appear:

- Professional but not intimidating
- Informed but not dogmatic.

Telephone communication forms a normal part of office practice, but should only be used when appropriate in relation to other forms of communication:

Advantages:
- Convenience

It is easier to use a telephone than to write and mail a letter
- Speed
Information can be relayed or requested immediately
- Cost
A telephone call in some cases may be cheaper than a typed letter
- Confidentiality
No permanent record remains of the conversation

Disadvantages:
- No permanent record
In some cases, a record may be useful or advisable. Memos can be taken, but are less reliable than a letter
- Cost
Long distance calls can be expensive, and quite often the contents of the conversation will have to be confirmed by letter in any event.

A tactless telephonist, poor telephone style, or failure to return calls can convey an unfavorable impression, and should be avoided.

b. Written Communication

When writing a letter, consider the purpose of the communication, which may be:

- To inform
- To record an event or conversation
- To request instructions
- To promote good relations.

The purpose of the letter and the nature of the recipient should affect the style of the letter (e.g., do not use obscure technical language when writing to a layperson).

Communication

Correspondence should be concise, relevant, and definite to be effective. If the contents are likely to exceed one paragraph, use separate paragraphs for each new point of importance. Numbering each point will help identification in replies and future reference.

End each letter on a positive note, summing up the reason for the communication (e.g., a request for information, details of the next action to be taken, etc.).

Letters should be checked before mailing for any typographical errors, spelling mistakes, or ambiguities which may convey a poor impression of the practice.

2. Promotional Literature

This may include:

- Brochures
- Newsletters
- Advertising.

Brochures

Many architectural firms produce brochures promoting their image and detailing their experience and achievements. The brochure may include:

- A statement of the firm's design philosophy
- Descriptions of areas of expertise (solar design, etc.)
- Details of past projects
- Details of the office and its personnel
- References (past clients, financial).

Again, careful consideration should be given to the production of the brochure, and how much time and money to spend on its production and distribution.

Loose-leaf binders may be useful to allow new data to be added without the necessity of reproduction of the entire brochure.

Newsletters

Some of the information used in brochures may be transmitted periodically by newsletters, which provide recent data about the office, its projects, personnel, and clientele.

Care should be taken in both cases not to violate any state regulations on advertising in the profession.

Advertising

Some firms may wish to advertise on a regular basis. Decisions to advertise should be made on the basis of state laws, the firm's requirements, and the principals' attitude toward the ethical considerations involved.

When considering the use of promotional literature, it may be advisable to utilize the services of a promotional agency. In all cases, consider first the cost-benefit issue before becoming involved in any such activities.

3. Public Involvement

It is not unusual for architects to be asked to give lectures, slide-shows, or interviews by community groups or the press. The benefits of this activity both to the profession and the individual office are many, but state requirements concerning advertising and publicity should not be violated.

Reference

AIA B-8, B-9.

The architect/owner relationship will be affected by the nature of the owner, who may be:

- A private individual
- A partnership
- A corporation or institution
- A state or federal department or governmental agency.

If dealing with employees of the larger organizations, their authority to bind the firm should be checked or verified at the outset.

In some cases, the owner and the user of the proposed project may not be the same party (as, for example, in the case of a school or hospital), and care must be taken not to confuse the requirements of the two roles.

The character of the owner will affect the administration of the project in a number of ways, including:

- Selection of the architect
- The architect/owner agreement
- Contractor selection procedures
- The owner/contractor agreement
- Methods of construction (see page 85)
- Means of communication between the respective organizations
- Forms and paperwork to be used.

State and federal agencies are likely to want to use their own forms and contracts, and will be bound by statutory requirements in the selection of both architects and contractors.

Selection of the Architect

This may be accomplished in three ways:

1. Directly:
   By reputation
   By recommendation
   By previous contact
   By chance
2. Comparatively:
   This method is usually used by institutions, public agencies, etc., where a number of architects will be asked to submit their resumés for consideration by a board. Information required may include:
   Age and achievements of the firm (examples of work, clever solutions, efficiency)
   Details of the practice (staff, workload, organization, and ability to take on new work)
   References (bank, former clients)
   Names of preferred consultants.

Interviews may also form part of this selection method.

3. Competitively
   Competitions may be:
   a. Selective
      Where a limited number of entrants will be invited to participate
   b. Open
      Where anyone may enter

Architects are advised only to enter competitions approved by the American Institute of Architects, and abide by the guidelines it has established.

The Agreement

The form of agreement between the architect and the owner is very important, and care should be taken to clearly establish the relationship at the outset. Oversights, omissions or misunderstand- ings at this stage may lead to serious problems later in the relationship which foresight and thorough attention will help to prevent.

There are a number of ways in which the architect/owner association can be formalized:

- By conduct of the parties (see page 73)
- By letter (see page 69)
- By formal written agreement
  A contract may be drafted for each new project, although the use of standardized forms is advisable. The AIA produces a number of standard forms (AIA Documents B141, B141/CM, B151, and B161; see page 105) although some owners, specifically larger institutions and governmental agencies, may wish to use their own standard forms. These should be studied carefully before signing.

Article Changes

In the eventuality that the standard contract articles have to be amended, omitted, or enlarged, great care should be taken to ensure that the terms of the agreement, as amended, do not adversely affect the architect's position with regard to liability, or conflict with provisions contained within related documents. If changes are necessary or required by the owner, legal counsel might be consulted to ensure that the overall documentation of the project remains consistent.

In most cases, the use of AIA forms is strongly advised, as they are generally accepted and understood throughout the building industry and are comprehensive in their coverage. Less formal methods of agreement may be used in projects of a simple or minor nature where a full contract appears inappropriate. Here, abbreviated contracts

The Architect/Owner Relationship

may be useful (AIA Document B151) or carefully drafted letters of agreement, which might include:

- Details of the extent and purpose of the project
- The general nature of the agreement
- Details of the site (location and address)
- The responsibilities and roles of each party
- Payment type and times of payment (see page 46)
- Details of retainers
- Methods of calculating fees and expenses
- Details of full and partial services
- Copyright considerations (see page 11)
- Additional services, if any
- Other matters (consultants, type of building contract, etc.).

Checklist

Factors to be considered at preliminary meetings, and possibly mentioned in letters of agreement and/or contracts include:

Obtaining details of:
- The owner and any representatives (names, addresses, etc.)
- The project (description of intent)
- The site
- The proposed user (if different from the owner)

Checking:
- The seriousness of the owner and ability to proceed with the work (even a credit check of some clients might be prudent at this stage)
- Whether any other architects are involved with the project (if so, they should be informed)
- The availability of office resources for the job
- Statutory requirements and consents necessary

Discussing:
- Appointment and payment of consultants
- Type of building contract to be used
- Method of contractor selection
- Single or separate contract system (see page 85)
- Early appointment of contractor
- Subcontractors and suppliers
- Methods of insurance and security (bonds, warranties)
- Limitation of liability and indemnities

Providing the owner with:
- Details of the owner/architect agreement and information, including details of payment
- Details of the architect to be in charge of the project
- Methods of communication
- Other data which will help the owner to understand respective duties and responsibilities and details of the construction process.

References

Sweet, pp. 73–84.
Walker, pp. 27–29.
AIA 11.

THE AMERICAN INSTITUTE OF ARCHITECTS

AIA Document B141

Standard Form of Agreement Between Owner and Architect

1977 EDITION

THIS DOCUMENT HAS IMPORTANT LEGAL CONSEQUENCES; CONSULTATION WITH AN ATTORNEY IS ENCOURAGED WITH RESPECT TO ITS COMPLETION OR MODIFICATION

AGREEMENT

made as of the FIFTEENTH day of MARCH in the year of Nineteen Hundred and EIGHTY ONE

BETWEEN the Owner: ELLEN I. WATER
P.O. BOX 314
HOLDEMAT BAY, WISCONSIN

and the Architect: FAIR & SQUARE A.I.A.
HOLDEMAT BAY, WISCONSIN

For the following Project:
(Include detailed description of Project location and scope.)

SINGLE FAMILY RESIDENCE TO BE LOCATED AT : 1 LAKESIDE, HOLDEMAT BAY,

WISCONSIN. THE 'WATER RESIDENCE' SHALL INCLUDE BUT NOT BE LIMITED TO ;

THREE (3) BEDROOMS, LIVING ROOM, DINING ROOM, KITCHEN, STUDIO, GREENHOUSE,

TWO CAR GARAGE.

The Owner and the Architect agree as set forth below.

The Architect's Services

The services which the architect may provide for the owner are categorized by the AIA Documents into two fields:

1. Basic services
2. Additional services.

1. Basic Services

These consist of five phases of work:

- Schematic design
- Design development
- Construction documents
- Bidding or negotiation
- Construction (administration of the construction contract).

At the completion of each phase, the architect should submit:

a. A statement of probable cost
b. A request for phase payment
c. A request for written authorization to proceed to the next phase.

2. Additional Services

Work undertaken by the architect which exceeds that stated under basic services should be identified and compensated in accordance with an agreed formula. The AIA Documents define additional services and provide for them to be paid for at a specified rate. Additional services may include:

- Additional site supervision (e.g., an on-site architect whose duties may extend to those of a construction manager)
- Programming
- Financial feasibility studies
- Planning surveys, site evaluations, and environmental studies
- Quantity surveys
- Energy studies
- Checking existing buildings or drawings, surveys, evaluations, etc.
- Preparation of alternative documents for bidding
- Detailed cost estimates
- Interior design, design of special furnishings
- Revision of drawings and specifications
- Drawings and specifications prepared in connection with certain change orders
- Evaluation of fire damage
- As-built drawings, presentation drawings, or models
- Expert witness, attendance at hearings
- Services necessary by default of contractor
- Post-completion inspections (see page 126)
- Post-occupancy evaluations
- Maintenance reports.

Phase Payments

Payment to the architect is usually made on a monthly basis reflecting the proportional allocation to each phase, with a down payment at the beginning of the agreement. The phases are usually divided into the following percentages for payment purposes:

- Schematic design—15%
- Design development—20%
- Construction documents—40%
- Bidding or negotiation—5%
- Construction—20%.

Types of Payment

The architect may be compensated by the owner in a number of ways:

- Stipulated or fixed sum
- Percentage of construction cost
- Multiple of direct personnel expense
- Professional fee plus expenses.

Stipulated or Fixed Sum

If the architect receives a fixed sum for all work undertaken in connection with the project, the owner is aware at the outset of his/her financial commitment in this respect. However, the architect must carefully estimate the work involved in order to determine a fair fee. If it is too high, the client may be dissuaded from proceeding; if too low, the architect may not make sufficient profit to render the work financially viable. The fixed sum method of payment should really only be used for small-scale work which can be accurately estimated, although some governmental agencies are constrained by law to require it.

Percentage of the Construction Cost

This is a common method of determining architectural fees in the private sector, although the owner might consider that this method provides the architect with little incentive for keeping overall costs down. Furthermore, a percentage of the final cost is not always reflective of the amount of work put in by the architect. If this method is chosen, the parties should take care to select an appropriate percentage rate.

Multiple of Direct Personnel Expense (MDPE)

By this method, the architect's fee is based upon the cost of each hour/day worked by principals and employees of the practice in connection with the project. This is multiplied by an agreed figure to allow for profit and overheads. It relates precisely to the architect's input in man-hours, and it is most useful where unknown or unpredictable circumstances might affect the work, such as:

- Unusual site conditions
- Unknown subsurface conditions
- New or experimental materials or building techniques
- Unusual building type
- Uncertain owner requirements
- Unforeseen circumstances (e.g., renovation of an old structure).

The MDPE method of payment necessitates the keeping of careful work records by the architect, and again gives no indication of the final cost, or of the architect's efforts to keep costs down, although it does remove the dependence of the architect's fee from the final cost of the project.

Professional Fee Plus Expenses

By this method, a fee is fixed at the beginning of the contract with the understanding that all expenses and costs will be reimbursed in addition to the agreed fee.

Reimbursable Expenses

Whatever payment type is used, the architect should be able to claim for expenses such as:

- Travel to and from the project
- Overnight accommodation and subsistence (if necessary)
- Long-distance communication
- Reproduction of data and postage
- Data processing and photographic work, if connected with additional services
- Overtime work and rates

- Renderings, models, etc.
- Additional insurance required by the owner.

In certain circumstances, it may be preferable to mix payment types to benefit both the architect and the owner. For example, on a complex renovation project, certain aspects requiring an unforeseeable amount of work (e.g., design, inspection, etc.) might be charged on a MDPE basis, whereas work of a more predictable nature (e.g., drawing and specification production) could be reimbursed on a fixed sum or percentage basis. In all cases, the architect should ensure that the agreement specifically provides for periodic payments.

References

Sweet, pp. 127–139.
Walker, pp. 22–23, 17–19.
AIA 11.

Action Required

1. Memo

MEMO

To: Bill
From: Tom
Date: Feb. 24, 1981
Re: New assistant

While you were out of town, I interviewed Hussein Chargeer for the open assistant position as agreed, and I am satisfied with his credentials.

I shall be at the AIA convention this week—could you draft a letter of appointment? A note on the terms discussed is on my desk, together with conditions of employment Xeroxed from the Personnel Policy Manual.

3. Letter

2/25/81

Dear Bill,

Sue-Ellen and I have decided to extend our place to provide a garage and a family room for the kids.

You being our neighbor, we would like you to draw up the plans. Will it cost a lot, and when could you start?

See you this weekend,

Yours,

Joe Kingly

5. Memo

MEMO

To: Tom Square
From: Bill Fair
Date: Feb. 25, 1981
Re: Possible work for Acme Corporation

A Mr. Sharp of Acme Corporation called. Says he wants a sketch design of a new office complex that you discussed with Acme recently by next Thursday, contract to follow. What's the story on this one?

2. Letter

WP/cg

February 24, 1981

Dear Sirs:

Further to our telephone conversation last week, we confirm our interest in employing you for the design of our new publishing offices in Euphoria, Iowa. We are anxious to get started as soon as possible, and would be pleased if you would send us the AIA contract documents that you suggested we use.

We look forward to hearing from you,

Yours faithfully,

Warren Pease

4. Letter

Our ref: EIW/jg   February 25, 1981

Dear Sirs:

I am thinking about having a new house built for my family in the Holdemat Bay area. Your practice has been recommended to me by a business acquaintance with whom I worked some years ago.

Could you send me some details of your practice and possibly some examples of your work for me to look at?

I look forward to hearing from you,

Yours faithfully,

Ellen I. Water

6. Letter from subcontractor

February 24, 1981

Dear Sirs:

We are sending you 150 carpet tiles (color green) for use in your main office. When we were there last, your technician mentioned how nice a new floor covering would be. Please accept these as a token of our continued satisfaction in working with your firm. We look forward to a similar working relationship in the future.

Yours sincerely,

Walter Wall

1. Letter

February 26, 1981

Our ref: BF/cc

Dear Mr. Chargeer:

Further to your recent discussion with my partner, Tom Square, on February 20, we are pleased to confirm your appointment as technician to this practice. As agreed, we would like you to begin your duties on May 1.

Details of your salary, terms of employment, working hours and holidays are enclosed on the attached sheets, which also provide information involving time off for illness, our pension scheme, contracting-out procedures and other office matters that you should know about. You are asked to read these carefully and retain them for future reference.

As a condition of your employment, we would require prior notice should you wish to undertake any private commissions.

I trust the foregoing is clear, but if you should be in doubt on any matter, please let us know. Tom and I look forward to working with you.

Yours sincerely,

Fair and Square

2. Memo

MEMO

To: B.F.
From: T.S.
Date: 2/26/81
Re: Project in Euphoria, Iowa

Before we start, check reciprocity between Wisconsin and Iowa—will we need to get a temporary license? If so, both of us will need one. We must be careful not to start work before we are registered in the state—could be grounds for nonpayment, so get licensing sorted out before we sign a contract and begin.

Also, not a bad idea at this stage to check out state and local laws (building codes, lien laws, etc.). A couple of phone calls to the AIA and Euphoria Building Department should do to start with.

3. Memo

MEMO

To: Tom
From: Bill
Date: Feb. 27, 1981
Re: Neighbor's extension

My neighbor wants us to design an extension for him. If you feel we need the work, I suggest an abbreviated form of agreement for a job of this size, or a well-worded letter. However, it's a bit close to home (my home!) for my liking, and we're pretty busy at the moment.

Would you mind if I suggested another firm because of our present full commitment?

4. Letter

Our ref: TS/cc
Your ref: EIW/jg

February 27, 1981

Dear Ms. Water:

Thank you for your letter of February 25 concerning a possible commission in Holdemat Bay. We have pleasure in enclosing our brochure which gives full details of our organization and our past projects. References may be provided upon request.

Should you decide to engage our services in this venture, we would be happy to meet with you to discuss the matter further.

If you have any further questions regarding our firm, please do not hesitate to contact us.

We look forward to hearing from you,

Yours sincerely,

Fair and Square

Diary insert

Re: Gifts, favors etc.
Date: 3/2/81

Have returned the carpet tiles with a note of thanks —as we're working with Walter Wall on the warehouse job. I have dealt with it immediately. The memo should clear things up in future, and should go into the Office Standards Manual for future reference. B.F.

5. Memo, diary insert and letter

MEMO

To: All employees
From: T.S. and B.F.
Re: GIFTS, GRATUITIES, AND FAVORS

It is the basic policy of this practice in the light of professional requirements of the State of Wisconsin not to accept any gifts, gratuities, or favors that could possibly be interpreted as causing a conflict of interest between our clients and ourselves.

All such offers to employees should therefore be returned to their senders to avoid any misunderstanding. A letter of appreciation should be included explaining the action, and thanking the donor for the generous thought.

Items of negligible cash value such as calendars, pens, etc. need not be returned. If there are any doubts about certain items, either of us should be consulted.

Your cooperation in this matter is greatly appreciated.

Letter

Our ref: BF/cc

March 2, 1981

Dear Mr. Wall:

Thank you so much for your offer of carpet tiles for the office. Unfortunately, our professional regulations do not allow us to accept gifts of this nature, and we are regretfully returning your kind gift.

Your thoughtfulness was most appreciated in the office, however, and we look forward to working with you again in the future.

Yours sincerely,

Fair and Square

6. Memo

MEMO

To: Tom
From: Bill
Date: 3/2/81
Re: Acme Corporation

Need more checking on this one. Are we dealing with Mr. Sharp personally or the Corporation? Have we established whether he has sufficient authority to form a contract with us in the latter case?

Maybe we could check out the Corporation (Better Business Bureau to start with), check Sharp's authority (in their Articles of Incorporation) and push for a firm contractual arrangement before we start—preferably the AIA one, but a letter might do.

I'll phone them this afternoon for written authority and ask for a retainer up front to test them out.

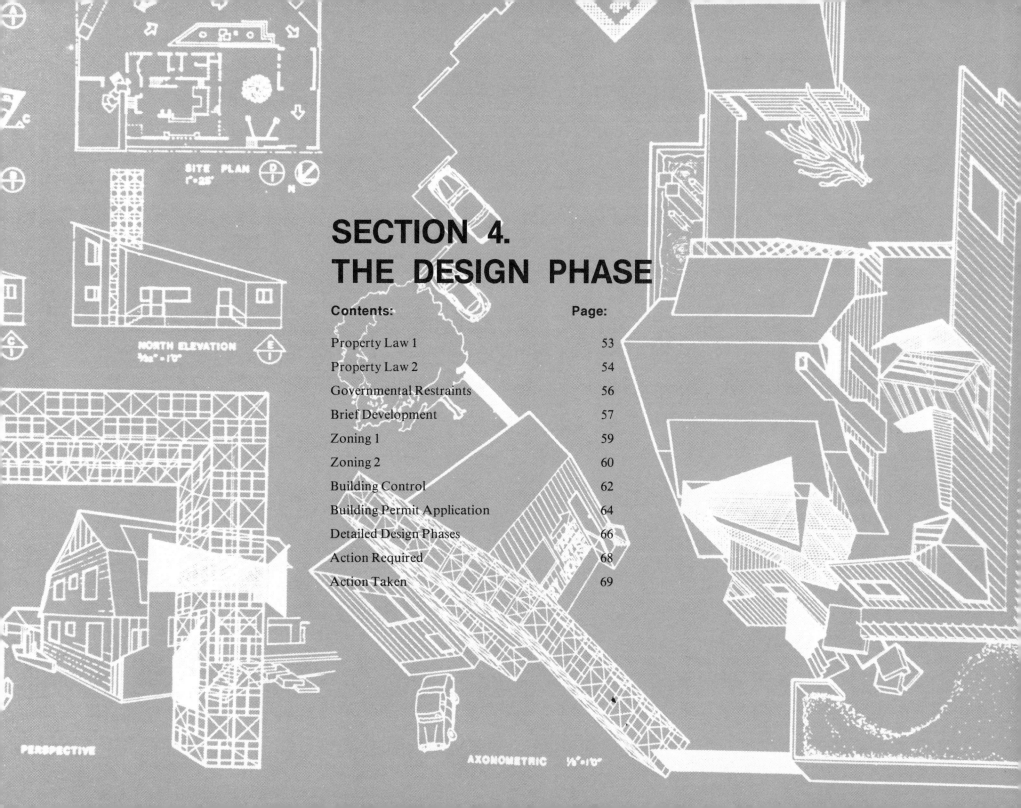

SECTION 4.
THE DESIGN PHASE

SITE PLAN
1"=25'
N

NORTH ELEVATION
1/32" = 1'0"

PERSPECTIVE

AXONOMETRIC   1/2"=1'0"

Certain legal rights, obligations, and restraints concerning land should be considered in the design process as they may affect:

- The choice of site for a particular development
- The character of the development, if the site has been selected
- The methods and procedures to be adopted in any proposed development.

Land law varies from state to state in its details and specific applications, but some general observations on a few important aspects of land law can be made.

Ownership

Ownership of land is expressed by *title,* which can be transferred from one party to another. Prior to purchase, a prospective owner normally has an investigation carried out into the background of the property:

- To ensure that the prospective seller actually possesses a transferable title to the land
- To check whether any encumbrances are attached to the title which might affect future enjoyment of the land
- To discover whether any governmental statutes or regulations exist which restrict development or usage of the land (see page 56).

Transfer of ownership is accomplished by *deed,* of which there are three basic types:

1. General warranty
2. Special warranty
3. Quitclaim.

1. General Warranty

By this kind of transfer, the transferor remains personally responsible for the title indefinitely. Although this is the most secure form of deed from the buyer's point of view, it is rarely granted.

2. Special Warranty

This guarantees that the land has not been encumbered during the current ownership, but gives no assurances in respect of the title prior to that period. This is the most common form of transfer deed, and its character requires that the title be carefully investigated.

3. Quitclaim

If a quitclaim is used, the owner promises nothing except that he/she will not contest the new ownership. This form of transfer is inadvisable in most circumstances.

Absolute and Acquired Rights in Land

Certain rights accrue to the owner of land which require no legal formality beyond the transfer deed (e.g., lateral support). Other rights can be of great importance, but must be formally acquired. These rights are often created by easements or convenants.

Easements and Covenants

These are legally enforceable, and attach certain conditions to specific land.

Easements are legal rights enjoyed by one party over the property of another. They are usually described in a deed or in a license, but in some circumstances, they can be implied by usage over a long period (e.g., 5-30 years continual use have been considered sufficient to imply an easement). Easements are frequently sought with regard to:

- Access
- Light
- View.

Restrictive covenants restrain an owner from undertaking certain actions in relation to his/her land. They are usually established by a previous owner (e.g., the developer, if the land forms part of a general development) and they are often introduced in an effort to protect the character of a neighborhood, or to maintain property values. Restrictive covenants may be used:

- To prevent fence building
- To assure minimum levels of aesthetic or architectural appearance
- To prevent major alteration or change to existing buildings
- To prevent tree-lopping.

Thorough investigation of the property title should reveal enforceable covenants of this nature. Some restrictive convenants are unconstitutional and therefore may not be enforced (e.g., racial restrictions).

In certain circumstances, easements can be sought by a new owner in connection with property of a neighbor (e.g., in order to dig out foundations close to the boundary). A license for this purpose should be requested by the new owners or their legal counsel, and may involve monetary consideration.

A prospective purchaser of land should ensure that a thorough check is carried out by a suitably qualified professional, to familiarize the buyer with any easements or covenants which may exist and which may affect the land's usage. This investigation should also include searches for other encumbrances (e.g., unpaid mortgages, outstanding liens, etc.).

Property Law 2

Other legal provisions exist which can affect the relationship between neighbors, and liability for persons entering upon the land. These include:

1. Spite fences
2. Tree ordinances
3. Nuisance
4. Occupier's liability.

1. Spite Fences

In some states, if a landowner maliciously erects a high fence which interferes with a neighbor's land (e.g., by causing excessive shading or view blocking), the courts can order the fence to be removed and allow the offended neighbor to claim damages. Other states refuse to interfere in the case of a spite fence on the basis that a landowner should have free use of the land owned. However, there appears to be a trend toward some control of spite fences.

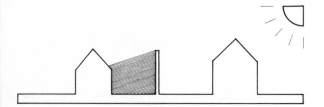

2. Tree Ordinances

In some localities, where a view is blocked by excessive foliage from a neighbor's tree, a reasonable request can be made to remove the obstruction, with costs to be shared between the two parties. Tree Commissions are sometimes set up to decide these matters in case of dispute.

3. Nuisance

Ownership of property generally entitles the owner to enjoy the land without interference by neighbors. Sometimes the activities of one party affect the enjoyment of the other to the extent that legal action might be taken to prevent further disturbance. An action for nuisance might be brought in respect of:

- Loud noises
- Anti-social activities (e.g., excessive vibration caused by pile-driving)
- Pungent odors or smoke
- Unsightly appearance of neighboring property.

It is important to remember that nuisance is classified as a tort, and in each case the court is concerned to discover whether a reasonable person would find the act complained of to be disturbing. The court will often also consider the conduct of the plaintiff, and the extent and duration of the alleged nuisance in reaching its decision. Another consideration is the benefit to society at large (e.g., industrial disturbances) which will be weighed against the disturbance caused to the plaintiff.

4. Occupier's Liability

Occupiers have a duty of care to all persons lawfully on their premises, and this duty varies according to the classification of the visitor.

a. Invitees

These are owed the highest duty of care by the occupier, who is responsible for those hazards known to exist, and those which could reasonably have been revealed. The category of invitees does not include social guests.

b. Licensees

A licensee is generally a person who comes onto the premises for personal reasons rather than for the purposes of the occupier, but with the occupier's consent (e.g., sales representatives). Social guests fall into this category. The occupier is obligated not to subject licensees to unreasonable risks, but this duty is reduced if licensees are in any way partially responsible for the injuries they sustain.

c. Passers-by

Boundaries should be clearly demarcated, and activities on the property conducted so as to show reasonable care in avoiding injury to passers-by.

d. Trespassers

Trespass may be defined as the unauthorized transgression of another person's land, including the air space above and the ground below. Trespass is classified as a tort (see page 4).

Even individuals who enter premises as trespassers are owed some duty of care, although this is reduced, particularly if the trespasser is intent on malicious damage. However, attempts to physically abuse an intruder should be limited to personal protection: courts have in the past held that protection of property alone does not justify extreme physical assault upon a trespasser, and in some cases, high damages have been awarded against the occupier.

The duty to child trespassers is generally higher than to adults because children may be less aware of property boundaries and inherent dangers than adults. Building sites, in particular, should be adequately secured and posted.

The above classifications are highly technical and it is difficult for ordinary persons to be sure of the status of all people who enter on their land. For this reason, occupiers should exercise great care in keeping their premises reasonably safe.

Reference

Sweet, pp. 191–199, 674–680.

Governmental Restraints

In the early stages of each project, attention should be paid to the various legal restrictions which might affect the scheme. Some restrictions apply to all projects, whereas others are applicable only to schemes of a certain type, or in a particular location.

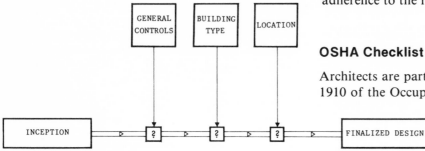

General Provisions

Although these vary according to state and locality, most building operations will require:

• A zoning permit (see pages 59–61)
• A building permit (see pages 62–65)
• Services connection (e.g., gas, electricity, water, sewage, telephone, etc.).

Special Provisions

In addition to the general requirements, some projects will require attention to:

• OSHA
• HUD.

OSHA

The Occupational Safety and Health Administration was set up by an Act of 1970 which makes it illegal to work in an unsafe place. OSHA has the authority to enforce safety standards, and to impose high financial penalties for violation of safety regulations. It enforces its standards by carrying out inspections of workplaces where accidents have occurred, or from which complaints have been made by employees. Spot checks are also made by OSHA officials to promote widespread adherence to the regulations.

OSHA Checklist

Architects are particularly concerned with section 1910 of the Occupational Safety and Health Act.

The following points may help to avoid problems with conformance to the OSHA regulations:

1. Obtain clear instructions of intended use from the client so that OSHA provisions can be considered in the early stages of design.

2. Ensure that the client is aware that compliance with OSHA can unavoidably increase the cost of the project.

3. Note that, in the case of a conflict between OSHA provisions and local building codes, the more stringent regulations prevail.

HUD

The Department of Housing and Urban Development was created in 1965 to help alleviate problems in urban areas by the promotion of major federal aid programs coupled with financial and technical assistance. If HUD is providing an input to a particular scheme, design guidelines may be laid down with which the architect is expected to comply.

Specific Types of Project

Particular projects are sometimes affected by specific legal constraints and need approval and/or inspection by individual state authorities which may impose their own standards (e.g., hospitals, schools, factories, etc.).

Location

Some restrictions affect all proposed projects within a specified area, particularly districts designated as historic preservation zones. In these, locally elected commissions develop and enforce rules and standards for future development. In some cases, individual buildings are singled out as historic landmarks, and this effectively prevents their demolition or alteration unless an appeal is successfully made according to the designated procedures (see page 60). Architects working in older districts or on the alteration or extension of older buildings should check first to discover whether the preceding restrictions have been imposed, and consider how the extra requirements will affect the design and progress of the project.

In assessing each project at the outset, the architect should ensure that the owner is aware of the scope of the architect's services in respect of gaining approvals so that there is no misunderstanding concerning the extent of basic services, and the possible need for additional payment (e.g., where a zoning appeal is necessary). It is most inadvisable for an architect to assure a client that the necessary approvals will be granted without difficulty. Attention to these matters at the early stages of the project will help to prevent any later decline in the architect/owner relationship.

Reference

Sweet, pp. 212–214.

Schematic Design Phase

Brief development is undertaken during the schematic design phase which is described in Article 1 of AIA Document B141 under basic services. During this phase, the architect is concerned to obtain a clear understanding of the owner's requirements. Following this, feasibility studies can be undertaken to ensure the owner's awareness of the viability of the project. If the owner approves continuation of the project, the architect will prepare schematic design drawings and other documents, and also submit an estimate of the probable cost of construction to the owner.

Preliminary Considerations

Factors affecting the project which may be given consideration at this stage include:

- Nature of the owner (public agency, private individual, etc.)
- Implications of ownership (see page 43)
- Type of building required (size, character, etc.)
- Degree of quality sought by the owner
- Any special conditions (e.g., future flexibility)
- Time available (for project completion)
- Finance available (for both the construction cost and the subsequent operational costs).

Effects

Although in some cases it may be too early to finalize decisions on the project, it is useful to consider the above matters together with their possible effects upon:

- Selection of the contractor (bidding or negotiation, see page 85)
- The method of bidding (see page 87)
- The type of contract

- Form of payment to the contractor
- Use of bonus/penalty clauses
- Use of construction manager (see page 28)
- Rate of liquidated damages
- Employment of consultants
- Insurance, bonds (see page 95)
- Use of separate or single contract systems (see page 85)
- Work required outside architect's basic services (see page 46).

Preliminary Procedures

Physical constraints, and their effect upon the owner's accommodation requirements will be clarified by the preparation of a site survey and accommodation schedule.

The owner's further requirements should be established by obtaining information on:

- Design objectives and criteria
- Possible constraints
- Space requirements and relationships
- Future flexibility and expandability
- Special equipment or systems required
- Landscape or site requirements.

Once all the necessary data is available, tentative schemes can be developed, incorporating:

- Location of proposed project on site
- Function and relationship of rooms and spaces (including their areas and heights)
- Primary elements (walls, floors, etc.)
- Overall appearance/character of the scheme.

Other more specific information which might be provided includes:

- Secondary elements (doors, windows, materials, special features)
- Integration of services (mechanical, electrical, etc.)
- Fixtures
- Other special requirements (e.g., special fittings, furniture, etc.).

To help the owner to decide on the best alternative available, feasibility studies may be undertaken and presented in the form of:

- A site plan
- Floor plans
- Sections
- Other (models, axonometrics, or sketches).

Before proceeding to the next design phase, the architect should ensure that written approval is obtained from the owner.

Surveys

According to AIA Document B141, Standard Form of Agreement between Owner and Architect, it is the owner's responsibility to provide all necessary descriptions of the site, and any further investigations which the architect considers appropriate. This may be done by employing the services of a land surveyor. However, the architect may prefer to undertake this work as part of additional services (see page 46).

Specific data should be provided to enable the architect to make an adequate assessment of the site, including:

- A legal description of the site
- Appraisal of existing structures on the site
- Site survey.

Brief Development

Site Surveys

These should provide the following information:

- Boundary delineation and demarcation
- Contour heights around the site, height of slabs, curbs, retaining walls, etc.
- Any encumbrances (e.g., rights of way, etc., see page 53)
- Dimensions and location of existing structures
- Details of any party walls
- Size, location, and species of trees
- Any special features
- Services (public and private, above and below ground e.g., sewers, manholes, etc.)
- Special topographical details and meteorological factors
- Roads and paths, both public and private, in and around the site (flow, usage, and direction, etc.)
- Points of entrance and exit
- Bench marks, etc.

Further information may be required depending upon the specific character of the project.

Note: Sketches and/or a photographic record of the site and surrounding area can prove useful for the architect as a ready reference source back at the office, particularly if the site is some distance away.

Although provision of site information is generally the contractual responsibility of the owner, the architect should be assured of its accuracy, as reliance on outdated or inaccurate data might be attributed to architectural negligence.

Statements of Probable Cost (AIA B141, Article 1)

At each stage in the basic services provided by the architect, an estimate of probable cost should be provided to the owner. This may be calculated by:

- Area and volume method
- Unit use method
- In-place unit method
- Quantity and cost method.

The method of cost calculation will depend upon the specific nature of the project.

References

Walker, p. 32.
AIA 11, 15.

Definitions

- *Zone:* to mark off into zones; specif., to divide (a city, etc.) into areas determined by specific restrictions on types of construction, as into residential and business areas.

(*Webster's New World Dictionary*)

- *Zoning Permit:* a permit issued by appropriate governmental authority authorizing land to be used for a specific purpose.

(*AIA Glossary of Construction Industry Terms*)

General

Zoning activity is the responsibility of individual states which pass zoning legislation as part of their police powers to protect the community. United States' zoning originated in the Enabling Acts of the 1920s which placed power to create and administer public land use regulations in the hands of local authorities. The most common type of zoning became known as Euclidean zoning; it consisted of establishing specific districts or zones for particular uses, e.g., commercial, manufacturing, residential, etc. These zones were then broken down into smaller units, e.g., light and heavy industry.

Scope

In addition to restricting use, the zoning regulations grew to cover matters such as:

- Density
- Light
- Air
- Space
- Height of buildings
- Bulk of buildings
- Plot sizes
- Aesthetic considerations.

Model Land Development Code

Although the Euclidean model was, and is, in common usage, cumulative zoning (i.e., allowing carefully regulated, multi-use districts) has gradually developed since the 1920s. In 1975, the Model Land Development Code was approved by the American Law Institute. The MLDC is only a recommended code, but some authorities have adopted some of its recommendations which include:

- Substantial responsibility for administering the development scheme should lie with local authorities.
- State authorities should provide some input, to avoid state problems resulting from purely local administration.
- Less rigid approaches to zoning.
- More stress on the environmental and aesthetic considerations of zoning.

Procedures

In many communities, the Building Department deals with zoning requirements, although where size and complexity of a community warrant a separate administration, zoning officers are sometimes appointed.

Applications for zoning and building permits are made simultaneously. If the applicant's request for a zoning permit is rejected, appeal procedures are generally available (see page 60). However, an appeal may not be the only alternative if a proposed project fails to match zoning requirements: procedures allowing greater zoning flexibility are available in many localities. These include:

1. Variances
2. Special use permits
3. Conditional permits
4. Rezoning.

1. Variances

Variances may be granted to enable land to be used for a different purpose than the category stated in the zoning ordinance. Specific requirements vary, but generally the applicant must show:

a. That exceptional circumstances exist
b. That strict application of the zoning ordinance would result in hardship
c. That the granting of the variance would not be detrimental to the public at large, or to those owning neighboring property.

Other restrictions often apply to the granting of variances.

2. Special Use Permits

Many localities make provision for the issuance of special use permits in given circumstances, which may be expressed in specific or general terms by the zoning ordinance.

3. Conditional Permits

In some cases, permission may be granted by the zoning authority contrary to the ordinance, provided that the applicant agrees to fulfill certain conditional requirements (e.g., noise control, provision of fences, etc.).

4. Rezoning

When an owner cannot match either the zoning requirements, or the conditions for a variance or a special use permit, application may be made to have the area in question rezoned. This is a difficult procedure especially if, as is often the case, neighboring landowners would suffer hardship as a result.

Common Features

Zoning is a complex and detailed field which can vary considerably from place to place. Care should be taken to gain an understanding of the zoning law which applies to the locality of a proposed project. However, there are several features common to many local zoning policies including:

1. Nonconforming uses
2. Floating and bonus zones
3. Environmental impact statements
4. Green belt/open space zoning
5. Conservation of historic buildings and landmarks.

1. Nonconforming Uses

Where an existing use failed to conform with new zoning segregation policy, the tendency was to eliminate that use. This approach caused considerable problems, and now several areas allow nonconforming uses to continue if they were lawful and existing prior to the new zone being established. Conditions for nonconforming uses vary from one locality to another.

2. Floating and Bonus Zones

Floating zones are sometimes located within specified zones to provide a measure of flexibility in future development. Bonus zones allow possible dispensation from the requirements of the zoning ordinance, provided that certain extras or "bonuses" are built into the project for the benefit of the community. For example, certain buildings in New York have been allowed to violate aspects of the ordinance on condition they provide public plazas or shopping arcades.

3. Environmental Impact Statements

These have been developed in some areas as a means of protecting and improving the environment by requiring detailed accounts of probable environmental consequences of certain zoning decisions. The statements are frequently particularly concerned with issues such as:

• Pollution
• Natural resources
• Coastlines and scenic features.

4. Green Belt/Open Space Zoning

This type of zoning is gaining support in the United States and it provides for the maintenance of open spaces, free from development and restricted to specific activities, e.g., recreation.

5. Conservation of Historic Buildings and Landmarks

Increasingly, local government authorities are taking steps to preserve historic districts, or individual historic buildings. Where this type of land use control exists, there are often provisions to alleviate the possible financial burden on the owner.

Appeals

A Board of Zoning Appeals is usually established in each locality and given the power to modify, reverse, or uphold zoning decisions. Appeals Boards are also usually empowered to grant variances and special or conditional permits. The duties and powers of the local Appeals Boards and the procedure which they must follow vary from place to place, but are generally defined in the relevant ordinances.

Appeal Procedure

Generally a Notice of Appeal must be made on the appropriate form which may require information and enclosures such as:

• Name and address of appellant
• Identification of property
• Name and address of agent (if any)
• Affidavit of appellant or agent
• Date of decision appealed
• Proposed use of property
• Present use of property
• Zoning classification
• Estimated cost of construction
• Copy of the decision against which the appeal is made
• Statement of grounds of appeal
• Certified plan survey
• Plans and drawings of the scheme
• Proof of ownership
• The requisite filing fee.

Other information and enclosures may be required, and some documents may have to be submitted with a specified number of copies. Generally, all zoning appeals must be made within the time limit stated in the relevant ordinance.

In the event that the appellant is unsuccessful in the appeal, there may be a further application to the regular courts in certain limited circumstances (e.g., if the Board of Zoning Appeals acted beyond the scope of its authority).

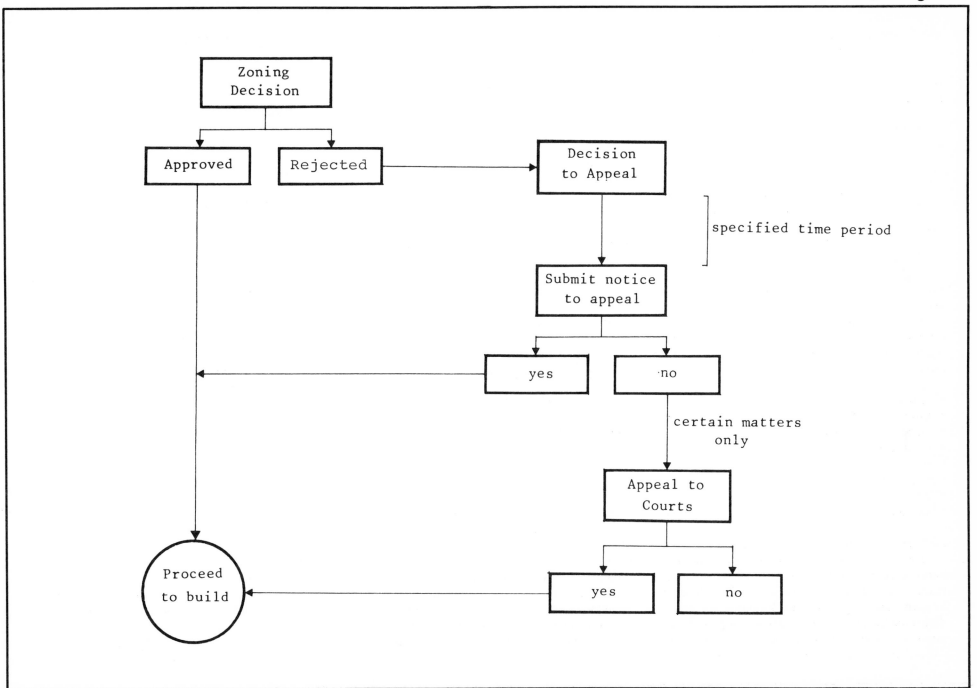

Building Control

Most construction projects have to conform to the requirements of the local building inspectorate, which are expressed in the form of a building code. Building codes vary in their scope and coverage, but tend to concern the following areas:

- Health
- Safety
- Welfare.

Typical codes include sanitary provisions, fire protection, structural requirements, etc. To ensure the code provisions are complied with, building permits are required for all building work (with a few exceptions; see page 64).

Buildings are divided into use or occupancy groups according to their proposed purpose. Each category has a separate set of requirements which must be matched in addition to the general provisions which apply to all building work. The occupancy groups may include:

- Assembly
- Business
- Factory and industrial
- Institutional
- High Hazard
- Mercantile
- Residential
- Storage.

Although several states have recently enacted statewide provisions for certain types of building, building control in the United States is predominantly carried out under the authority delegated by each state to the individual locality. This means that each local authority can develop and administer its own individual code, which generally consists of several sections, dealing with:

- Building
- Electrics
- Plumbing
- Heating, ventilation, and refrigeration
- Housing
- Fire.

In addition, most localities have separate Fire Prevention Codes which are administered by the local Fire Department.

Types of Code

Building code requirements can be expressed in different ways as:

1. Specific regulations
2. Functional requirements
3. Performance Standards.

1. Specific Regulations

These are basic statements of a direct nature, giving a fixed and clear standard providing a limited range of solutions and little flexibility of choice.

2. Functional Requirements

These give complete freedom to provide a solution by making very generalized requirements, without indicating how the desired level of performance might be achieved (e.g., buildings should be designed and constructed so that if a fire breaks out, everybody can evacuate the building and immediate area in safety). Despite their flexibility, functional requirements are sometimes criticized because of the lack of direction in their demands.

3. Performance Standards

These provide an intermediate alternative to the other forms of regulation by providing a measurable and precise account of the performance that is required, but leaving it to the designer to decide exactly how to comply. Performance standards are favored by most of the model building codes.

The Model Codes

In an effort to assist local authorities in the development of building codes which are efficient and up-to-date, several model codes have been published by groups concerned with the building control process, including:

- The Uniform Building Code, published by the International Conference of Building Officials (ICBO)
- The Basic Building Code, published by the Building Officials' Conference of America (BOCA)
- The National Building Code, published by the American Insurance Association
- The Southern Standard Building Code, published by the Southern Building Code Congress International.

Specialist model codes have also been developed such as:

- Fire Codes (e.g., the Basic Fire Code published by BOCA)
- Electrical codes (e.g., the National Electrical Code published by the National Fire Prevention Association (NFPA))
- Plumbing codes (e.g., the Basic Plumbing Code published by BOCA).

The model codes are widely used in the regulatory process, but owing to their voluntary status, each authority may amend, alter, or ignore the models as they wish. Out-of-date provisions have also been identified as a problem. Architects working over a wide geographic area should take care in checking the regulatory requirements specific to each project during the early design stages and, in the event of uncertainty, contact the relevant Building Department.

Standards

The building codes are a tabulation of accepted standards in building practice, and are developed by reference to organizations which test and give official approval to new building materials and techniques. The opinions of each organization are not binding upon local authorities, but the more respected of these tend to be widely accepted, including:

- The American Society for Testing and Materials (ASTM)
- The National Fire Protection Association (NFPA)
- The American National Standards Institute Inc. (ANSI)
- United States Department of Commerce (USDC).

Building Permit Application

Each locality administers its own building regulations and specifies the procedures which applicants must follow to obtain a permit. Outlined here are some common aspects of the procedures followed in many localities, but architects and developers should take care to discover the application procedures relevant to the locality of the project. Any questions should be directed to the relevant Building Department. Architects require written authority if they intend to make a permit application as the landowner's agent.

Need for Permit

Most building work requires a permit including:

- Construction
- Demolition
- Occupancy
- Heating and cooling installation
- Moving of buildings
- Alterations.

Exemptions are limited to minor alterations and ordinary repairs where:

1. No work of a structural nature is proposed, and
2. Health and safety are not affected.

Applications for permits usually consist of the following:

- The completed form of application which includes a description of the work, location, occupancy use, and other information required by the individual building department
- A plot survey showing proposed and existing buildings, distances to lot lines, and accurate boundary line information, etc.

- Building plans including drawings and specifications of sufficient size and detail to show the character and nature of the work (foundation plans, floor and roof plans showing exits, etc.)
- Additional details, e.g., structural, mechanical, electrical drawings, computations, stress diagrams, etc. Structural calculations and provisions for fire resistance.
- The required fees.

The building inspectorate will consider the application and either:

- Issue a building permit
- Reject the application in writing, specifying reasons for the rejection.

If a permit is issued, work must generally begin within a fixed period and be completed within a specified time.

A notice signed by the building inspector confirming the issuance of the permit must generally be displayed on the construction site.

The building inspector has the right to enter the site at any reasonable time to check that the work is in compliance with the code. Required inspections are likely to be undertaken at important stages in the construction process, and due notice must be given to the inspector before such work begins. These stages may include:

- Foundations (trench, reinforcement, weatherproofing, etc.)
- Concrete slabs and framing
- Roofing
- Electrical work
- Gas piping and fixtures
- Plumbing and sewer connections

- Heating, ventilation, and refrigeration
- Plastering (interior and exterior).

The costs of any tests which are ordered by the building inspector in connection with these duties must be borne by the owner.

Certificate of Occupancy

When the work is completed, a certificate of occupancy must usually be obtained before the building can be used. This certificate confirms that all the building regulations have been complied with, and must be available for inspection on the premises at all times.

If a part or portion of the work is ready for occupation before the completion of the project as a whole, a temporary certificate may be issued for the part of the work concerned.

Variations

Where there are practical difficulties in carrying out the requirements of the building codes, the inspectorate may, upon written request, vary or modify them as long as the spirit of the law is upheld, and health and safety provisions are not affected.

Violations

If at any time during the construction process the inspector feels that the work is not sufficient to satisfy the building code, the contractor will be required to amend the work before further certification. In certain instances, the building permit can be revoked, and a *stop work order* may be issued if the inspector considers the violation to be sufficiently serious.

Appeals

If an application for a building permit or a variation is rejected, there is generally provision for appeal to a Board of Appeals which is empowered to uphold, reverse, or modify the inspector's decision.

If the appellant is not content with the Board's ruling, a court action might be considered if there are sufficient grounds. Advice of legal counsel should be sought before such action is taken.

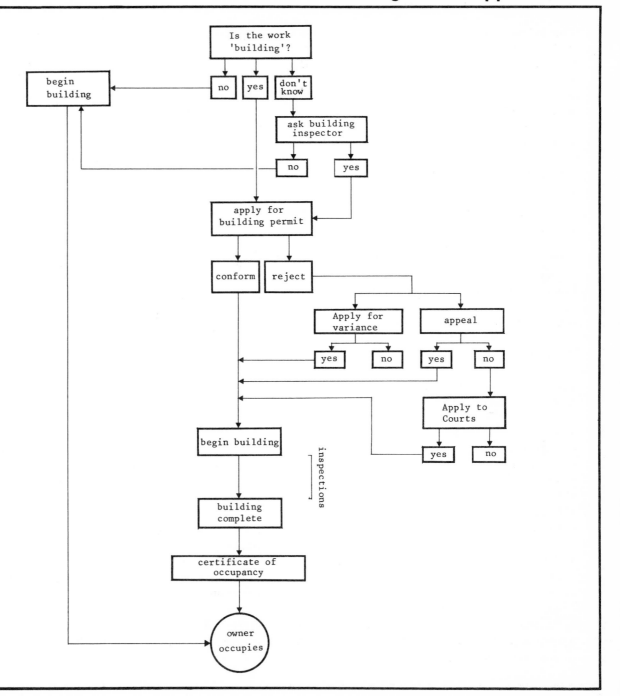

Detailed Design Phases

The architect's basic services beyond the schematic design phase (see page 57) and prior to the commencement of construction are stated in the AIA contract documents as:

- Design development phase
- Construction documents phase
- Bidding or negotiation phase (see page 87).

Design Development

Upon receiving written authority by the owner to proceed, the architect should start preparing more detailed illustrations and data relating to the proposed design. Any consultants which have been employed may give assistance at this stage, providing integrated input into the design process so that a final scheme can be prepared for the owner's approval, ready for the construction documents phase to begin. A further statement of probable cost must also be submitted to the owner during this phase.

Submittals to the owner concerning the development of the design could include the following:

- Site plan
- Floor plans
- Elevations
- Sections
- Schedules and notes
- Calculations
- Preliminary draft of the project manual
- Outline specifications
- Other data (e.g., electrical and mechanical systems).

The importance and significance of decisions made at this stage should be made clear to the owner, and written approval to proceed obtained before continuing to the next phase.

The Project Manual

During the design development phase, a project manual should be drafted. This will contain the bidding requirements and the contract documents, including the technical specifications, but excluding the drawings.

Construction Documents Phase

When details of the project have been sufficiently determined and approved by the owner, the architect will undertake:

- Preparation of detailed working drawings and specifications sufficient for construction purposes
- Assistance to the owner in securing bidding information, forms, contracts, and conditions (see page 87)
- Any further changes in the statement of probable construction cost
- Assistance to the owner in filing for any government approvals (see page 56).

Drawings

Some offices develop standardized practices in respect of working drawings and the construction documentation phase. These may include the following:

At the beginning of the documentation phase, estimate the number and type of drawings necessary to complement the specifications, and prepare a drawings schedule. This will facilitate office programming and improve the production sequence.

Draw only as much as is necessary. Time and money are often wasted in the duplication of material which is adequately covered elsewhere (e.g., in the specifications or schedules). Again, careful planning at the outset can help to prevent inefficiencies. Generally, information that relates to quantity and location should be in the drawings, while that pertaining to quality, method, and result should be in the specifications.

Consider whether time can be saved in the actual drawing process. Allocate simpler tasks to junior employees, use base negatives to add "layers" of information onto duplicated negatives or use a light table to trace basic repetitive work and/or check for consistency on the different layers.

Use standard methods of cross referencing throughout all projects so that employees become familiar with their use.

Although the number and mix of drawings will vary from project to project, the basic range of drawings is as follows:

Key drawings:
- site plan
- floor and roof plan
- elevations
- sections
- details
- schedules (e.g., doors, windows)

Structure and assembly drawings:
- foundation layout
- floor and roof layouts
- sectional structural details
- relevant schedules (e.g., columns, beams)

Mechanical drawings:
- mechanical layouts
- plumbing data and schedules

Electrical drawings:

- heating, ventilating, and air conditioning layouts and data
- stacks
- electrical layouts
- electrical details and schedules

Use standard size tracing paper for ease of storage, collating, etc. Smaller sheets could be cut in proportion to a base sheet for consistency.

Make sure each sheet contains:

- Project number
- Project title and location
- Sheet number and title
- Scale
- Drafter's name
- Checker's name
- Date (of issuance and revision)
- North point (where relevant)
- Space for stamp
- Revision space
- The name and address of the practice
- Space for owner's approval signature.

Build up a collection of standard details or schedules that may be used or traced in future schemes.

Specifications

Because of their important interrelationship, drawings and specifications should be developed together to avoid any duplication or omission of information. In accordance with the Uniform System for Construction Specifications, Data Filing and Cost Accounting (AIA Document K103), specifications are split into sixteen parts, or divisions.

The MASTERSPEC system may also be useful to the architect in the preparation of specifications: this is a computerized resource based upon the Uniform System and, at a time when it is becoming increasingly difficult to keep up to date with technical developments, represents a comprehensive and current professional aid.

Great care should be taken in the preparation of the specifications, which should generally either be handled by one of the principals or a specialist employed specifically for the purpose.

Specifications should be as concise and comprehensive as possible to prevent the necessity of too many addenda and/or the passing of alternates or unit prices (see page 87). Furthermore, the use of specific trade names should be avoided where possible.

References

Sweet, pp. 144–161, 313–316.
Walker, pp. 12–15.
Acret, p. 69.
AIA 11, 12, 14.

Action Required

1. Letter

March 10, 1981

Our ref: EIW/jg
Your ref: TS/cc

Dear Sirs,

Thank you for your letter of Febraury 27 and the enclosed brochure. I am satisfied that your firm can provide the services I require, and would like to meet with you to discuss the scheme and the contractual arrangements in the near future.

Please phone my secretary to set up an appointment.

Yours faithfully,

Ellen I. Water

2. Letter

March 10, 1981

Dear Sirs,

I am writing about the rejection of my proposed Bowling Center in Holdemat Bay by the building authorities. On reflection, I feel that it is worthwhile pursuing, and would like to appeal the decision.

If we made a really good presentation—models, photographs, etc.—I thing it would put the scheme in a better light. Do you think this is possible, and would it take you long to do the work?

Sincerely,

A. Gayne

3. Memo

MEMO

To: Bill
From: Tom
Date: 3/11/81
Re: Lending out stamp

Our old colleague, Dick, at Designer's Ink phoned today. He is only a few months from qualification now, and needs an architect's seal on a project he's working on—wants to know if we would put ours on for old times sake, and assures us the project is absolutely checked out. What do you think?

4. Letter

March 11, 1981

Dear Sir,

I am writing to complain about your men on the house being built next to mine at Witt's End. For a start, they have dumped a load of bricks and steel over part of my yard, blocking the gateway. Also, the noise and dust caused by the digging is making my kids sick, and I want it stopped.

If you do not act immediately to deal with these nuisances, I will be forced to take legal action.

Yours,

Vic Sassious.

1. Letter and diary insert

Our ref: TS/cc
Your ref: ElW/jg

March 12, 1981

Dear Ms. Water:
    re: New house at Holdemat Bay, Wisconsin

We are happy to confirm your verbal decision made during our meeting this morning to employ the firm of Fair and Square as architects on the above project.

As agreed during our discussion, all work connected with the construction of your house will be carried out on a stipulated sum basis and be competitively bid.

Our compensation for architectural services will be calculated at x% of the final cost of the construction work as agreed, and paid at periodic intervals during the project. Details of this and our respective rights, duties, and obligations are fully covered in the attached AIA Document B141, "Terms and Conditions of Agreement between Owner and Architect," which will form part of the agreement between us.

Letter

2.

Any additional services that you may require, as specified in the enclosed document, will be undertaken upon your written authorization and be charged on the basis of x times the direct personnel expense. This will be calculated at the usual rates of pay plus benefits of personnel and will include the principal's time calculated at $x per hour.

Please sign the enclosed copy of this letter and return it to our office so that we can proceed with the design work without further delay.

We look forward to working with you on this project. If there is any further information that you require, please do not hesitate to contact us.

        Yours sincerely,

        Fair and Square

---

3. Memo

MEMO

To: Tom
From: Bill
Date: 3/12/81
Re: Lending out seal.

NO! State codes prohibit it and we would be crazy to take full responsibility for work which
    a) we don't control
    b) we don't get paid for.

I'll drop Dick a line explaining our predicament.

Diary insert

Date: March 12, 1981
Re: Ellen I. Water project.

Discussion with Ellen I. Water this morning. All details discussed, and a letter of agreement ready to go out.

---

2. Letter

Our ref: BF/cc

March 13, 1981

Dear Mr. Gayne:
    re: Proposed Bowling Center,
       Holdemat Bay, Wisconsin

Thank you for your letter of March 10 enquiring about appealing the building authorities' decision regarding the above project. There is an appeals procedure in Holdemat Bay, and a hearing may be requested within 10 days of the decision. This does not give us much time, but, upon your written request, we will proceed with an appropriate presentation of the scheme for appeal purposes.

Additional payment in respect of this work and attendance at the hearing will be charged according to Article 5 of our agreement.

We look forward to hearing from you,

        Yours sincerely,

---

4. Letter and diary insert

Letter

Our ref: BF/cc

March 13, 1981

Dear Mr. Sassious:
    Re: Project at Witt's End, Wisconsin

Thank you for your letter of March 11, the contents of which we note.

As you will appreciate, although we are the architects in respect of the above project, we have no authority in the areas you mention, and are unable to comment on the matter. However, we have promptly forwarded your letter to the contractor in charge of the site.

        Yours sincerely,

        Fair and Square

Diary insert

Date: March 13
Re: Vic Sassious' letter on trespass and nuisance.

Have replied to Sassious, saying not our responsibility. Better send his note on to the contractor and warn him—could be a claim for trespass and possible nuisance. Not our concern really, but we want to make sure that the job doesn't get delayed at all.

SECTION 5.
CONTRACT FORMATION

Contents

A contract has been defined as:

"A legally binding agreement between two or more parties, by which rights are acquired by one or more to acts or forebearances on the part of the other or others"

Sir William Anson

Formation

Contracts may be formed in a number of ways:

a. Orally

A contractual relationship may be formed between parties in some instances where no written agreement exists, but a verbal contract was made.

b. By Conduct

The actions of parties may be such as to imply a contractual relationship between them.

c. In Writing

Certain kinds of contracts should be formed in writing if they are to be enforceable (e.g., in some states, real estate brokerage service contracts must be in writing).

d. Under Seal

Contracts made in this form traditionally did not require consideration to be enforceable in the courts (see below). However, the law relating to the status of sealed contracts now differs greatly from state to state.

e. Evidenced in Writing

Some contracts must be evidenced in writing if they are to be enforceable (e.g., contracts which, according to their terms, cannot be performed within one year).

Status

Contracts may be:

- Valid
- Void—without any legal effect
- Voidable—i.e., valid until one of the parties repudiates
- Unenforceable—in the courts.

Elements of a Valid Contract

There are a number of basic elements which are necessary for the creation of a legally binding and enforceable contract. These are:

1. Offer and Acceptance

An offer by one party must be clearly made, and that offer must be unconditionally accepted by the other party or parties. Upon acceptance, the contract comes into effect.

2. Intention

Intention must be shown by all parties to enter into a binding contract.

3. Capacity

All parties to the contract must be legally capable. For example, minors and persons of unsound mind may be excluded from certain types of contract.

4. Consent

Consent must be proper, and not obtained by fraud or duress.

5. Legality

The contract must be formed within the boundaries of the law. For example, a contract to commit a crime would not be binding.

6. Possibility

Contracts formed to undertake impossible tasks are unenforceable.

7. Consideration

Each party must contribute something in consideration of the other's promise. Consideration must be:

- Real
- Legal
- Not necessarily adequate
- Possible
- Not in the past
- And must move from the promisee.

Privity

Privity is a legal doctrine which recognizes that only a party to a contract may sue upon it. There are certain exceptions to this general rule, e.g., where an agency relationship exists, the principal is bound by contracts entered into by his agent with third parties.

Contract Law

Discharge

A contract may be discharged by:

- Performance, i.e., realization of the agreement within the terms of the contract.
- Agreement by all parties to cease their contractual relationship.
- Operation of law, e.g., if a contract is made for a limited period, and that period expires.
- Frustration or subsequent impossibility. Performance of the contract may be possible at the outset, but later be frustrated by events (e.g., death of a party, destruction of an element constituting the basis of the contract).

Breach

A breach occurs when a party to the contract does not fulfill his/her obligations. If the breach "goes to the root" of the agreement, the contract is treated as discharged. Such breaches are referred to as "material," and the injured party may seek one of the following remedies:

1. Refusal of Further Performance of the Contract

2. Rescission

This is a discretionary remedy, enabling the courts to cancel or annul the contract.

3. An Action for Specific Performance

If successful, the court orders the party in breach to fulfill his/her obligations under the contract.

4. An Action for an Injunction

An injunction is a legal means of preventing further action by the party in breach.

5. An Action for Damages

This is the most common remedy for breach of contract. Damages can be:

- General, i.e., arising out of the breach
- Nominal, if the breach is merely technical
- Punitive or exemplary, if the court considers the defendant's behavior particularly deplorable
- Liquidated, i.e., ascertained, as in most construction contracts
- Unliquidated, i.e., unascertained.

6. An Action for a Quantum Meruit

This is a claim for an amount equal to that which the plaintiff has earned in respect of the contract.

Types of Contract

A building contract may take any form which is agreeable to the parties involved, but certain proven types of contract have been developed which are useful for certain building projects. These are:

1. Fixed price/stipulated sum contracts
2. Cost-type contracts.

1. Fixed Price/Stipulated Sum

This method of agreeing a contract price is widely used in the construction industry, where one party pays an acceptable sum for a specified amount of work to another party who agrees to undertake it. It is nearly always used in connection with competitively bid work (some public bodies are constrained by law to use this method), and has the advantages of:

- Enabling the owner to know the final cost of construction at the outset of the work
- Releasing the contractor from having to keep accurate time records for the owner's scrutiny.

Stipulated sum contracts have certain disadvantages: for example, escalation or inflation of prices or unforeseen circumstances might affect the contractor's fixed profit margin. In some cases, this could mean a higher base bid to cover such contingencies, and so the owner may pay more than is strictly necessary. However, standard forms of contract often include equitable clauses to deal with these matters (e.g., escalation clauses, concealed conditions clauses, etc.). The fixed price method of contracting is most suited to building projects of a predictable nature, where a full set of construction documents is available. In federal projects a fixed price-incentive firm method of contracting has been developed.

2. Cost-Type Contracts

In this type of contract, the owner reimburses the contractor for the actual cost of completing the work, together with a negotiated fee. The fee might be:

- A percentage of the final construction cost
- A fixed fee
- An award fee.

The method is useful:

- In negotiated contracts
- Where unknown conditions might be encountered
- If new building methods/materials are being used
- Where final scheme designs are not fully completed.

It has the advantage of not needing a full set of documents before a price can be negotiated and work started, and enables the owner to bring the contractor into the process at an earlier stage, if required. It may be disadvantageous in that the owner is uncertain of the final cost and, in its percentage form, the cost-type contract gives no incentive to the contractor to keep costs down. Furthermore, the contractor will be obliged to keep accurate records of the work for payment purposes. In fact, cost-type contracts are prohibited in some states for certain kinds of work.

Certain variations upon the cost-type contract have been developed which make it more viable. These are:

a. Cost plus award fee
b. Cost plus incentive fee.

a. Cost Plus Award Fee (CPAF Contract)

This is often used in federal procurement work where the fee is negotiated on a percentage of the agreed estimated final cost. Added to this will then be an award fee, which may be two or three times the base fee, and is paid by the owner upon previously established criteria.

b. Cost Plus Incentive Fee (CPIF Contract)

By this system, the contracting parties negotiate a target cost and fee, a base and ceiling fee, and a fee adjustment formula to provide an incentive for early and economic completion.

Foreign Contracts

Architects may often be called upon to work in other states or, increasingly, other countries. As well as considering licensing requirements (see pages 48–49), the architect should take great care at the contract formation stage to avoid difficulties which might arise in enforcing agreements due to:

- Conflicts between state laws (e.g., lien laws)
- Conflicts between national laws
- The contractual capacity or immunity of parties operating in other states or countries.

Before any major agreements with a foreign element are entered into, it would be prudent to check the legal position with legal counsel.

Contract Checklist

Some of the more important factors to be considered when determining the type of building contract are:

Building Contracts

- Type of project (unusual building, renovation, etc.)
- Method of construction proposed (experimental techniques, etc.)
- Size and complexity of project
- Time constraints
- Finance available
- Degree of certainty of the owner's requirements
- Progress of the construction documents
- Probability of further changes
- Amount of information available at contract formation
- Availability/desirability of accurate cost prediction
- Expertise necessary/available
- External factors or problems (e.g., site constraints, labor shortages, etc.)
- Quality of work required (luxury, low cost, etc.)

References

Sweet, pp. 241–255.
AIA 16, 17, 18.

Just as any type of contract can be selected by the parties involved, so any form of agreement can be used to determine the terms and conditions of the contractual relationship.

However, in the interests of both parties, it is generally recognized that a format which has common usage and understanding is preferable. Standard forms of contract have been developed by a number of bodies, including:

- Professional associations
- The federal government
- State agencies
- Large institutions.

In some cases, the use of house forms is a requirement of the owner. These should be carefully scrutinized, particularly in relation to the owner-architect agreement to ensure that provisions affecting the architect's duties do not violate state licensing requirements or increase the architect's liability.

Where possible, AIA contract documents should be recommended. These have been developed over a long period of time, and are recognized throughout the construction industry. Furthermore, they may be used in conjunction with a wide range of other AIA standardized documents (see page 105).

AIA Forms

The major standard forms produced by the AIA for building purposes are:

- A101, Owner-Contractor Agreement, (Stipulated Sum)
- A101/CM, Owner-Contractor Agreement, (Stipulated Sum), Construction Management Edition

- A107, Short Form for Small Construction Contracts, (Stipulated Sum)
- A111, Owner-Contractor Agreement, (Cost Plus Fee)
- A117, Abbreviated Owner-Contractor Agreement, (Cost Plus Fee)
- A171, Owner-Interiors Contractor Agreement
- A177, Abbreviated Owner-Contractor Agreement for Furniture, Furnishings, and Equipment.

These basic forms of agreement are supplemented with the *general conditions* of the contract, which provide details of both parties' duties and responsibilities, and provide mechanisms for action in the event of certain circumstances (e.g., possible changes, etc.):

- A201, General Conditions of the Contract for Construction (see pages 79-84 for commentary)
- A201/CM, General Conditions of the Contract for Construction, Construction Management Edition
- A201/SC, General Conditions of the Contract for Construction and Federal Supplementary Conditions
- A271, General Conditions of the Contract for Furniture, Furnishings, and Equipment.

Supplementary Conditions

Despite the broad and thorough coverage of the standard forms of contract, it is not unusual for variations or additions to be required. These may occur through:

- Owner's requirements
- The nature of the project
- Local/state legal requirements
- Climatic or physical factors.

Where variations to the general conditions are necessary or required, they may be included:

- In the instructions to bidders
- In the owner-contractor agreement
- By modification of the general conditions of the contract
- By the addition of the AIA Supplementary Conditions
- By inclusion in the General Requirements of the Specifications (Division 1, Uniform Construction Index).

Location of new addenda usually follows these generally accepted principles:

- Anything concerning the bidding phase and not affecting the construction phase should be included in the Instructions to Bidders
- Matters going to the root of the contract (price, time, etc.) should be included in the owner-contractor agreement
- Legal matters which may vary according to location (indemnification, insurance, etc.) should be dealt with in the Supplementary Conditions
- Matters of a procedural or administrative nature (e.g., temporary structures, etc.) should be included in the General Requirements (Division 1) of the Specifications.

Great care should be taken in the reformulation of contract documents if any changes are anticipated, and certain matters (e.g., legal responsibilities) should not be attempted without professional legal assistance.

Standard Form of Agreement

THE AMERICAN INSTITUTE OF ARCHITECTS

AIA Document A101

Standard Form of Agreement Between Owner and Contractor

where the basis of payment is a

STIPULATED SUM

1977 EDITION

THIS DOCUMENT HAS IMPORTANT LEGAL CONSEQUENCES; CONSULTATION WITH AN ATTORNEY IS ENCOURAGED WITH RESPECT TO ITS COMPLETION OR MODIFICATION

Use only with the 1976 Edition of AIA Document A201, General Conditions of the Contract for Construction.

This document has been approved and endorsed by The Associated General Contractors of America.

AGREEMENT

made as of the FIFTEENTH   day of MAY   in the year of Nineteen Hundred and EIGHTY ONE

BETWEEN the Owner: ELLEN I. WATER
P.O. BOX 314
HOLDEMAT BAY, WISCONSIN

and the Contractor: PHILIP DA TRENCHTYN INK.
WITTS END
WISCONSIN

The Project: WATER RESIDENCE, 1 LAKESIDE, HOLDEMAT BAY, WISCONSIN

The Architect: FAIR & SQUARE A.I.A.
HOLDEMAT BAY, WISCONSIN

The Owner and the Contractor agree as set forth below.

AIA DOCUMENT A101 • OWNER-CONTRACTOR AGREEMENT • ELEVENTH EDITION • JUNE 1977 • AIA® ©1977 • THE AMERICAN INSTITUTE OF ARCHITECTS, 1735 NEW YORK AVE. N.W., WASHINGTON, D. C. 20006   **A101-1977** **1**

(page I of 4)

ARTICLE/PARAGRAPH/SUBPARA	CONTENT	ARCHITECT'S POWERS/DUTIES	COMMENT
ARTICLE 1	CONTRACT DOCUMENTS		
1.1.2.	The Contract	Architect is entitled to performance of obligations which the contract intends for the architect's benefit	The Contract Documents create no contractual relationship between the architect & contractor (or subcontractors) or the architect/owner
1.2.1.	Execution, Correlation, and Intent	Check all Contract Documents signed in triplicate by owner & contractor If not signed, the architect shall identify the contract documents	
1.3.1.	Ownership and Use of Documents	Check all documents returned after completion (except one contract set per party)	Drawings and specifications are the property of the architect and cannot be used for any other purpose than the project to which the contract refers
ARTICLE 2	ARCHITECT		
2.2.1.	Administration of the Contract		The architect will provide "administration of the contract"
2.2.2.		Check owner aware of all *his* duties and pass on all communications. Do only what is required in contract.	The architect will 'advise and consult' with the owner & act *only to extent provided in the Contract.* The owner will pass all communications to the contractor through the architect
2.2.3.		Visit site at "appropriate" intervals (stage of work, type of job, etc.) to check progress & quality. Keep owner informed.	No need to make continual site inspections
2.2.4.		Check conformity of the work against the contract documents under 2.2.13.	Architect not responsible for construction means, safety on site or failure of contractor to carry out work properly.
2.2.5.		The architect has right of access to the site, contractor's workshops, etc., if relevant to project.	
2.2.6.		Determine amounts due to contractor and issue certificates (see 9.4.)	
2.2.7.		*Interpret* the contract documents and judge the performance of the work, and compliance with contract.	
2.2.8.		Respond with "reasonable promptness" to *written* requests for interpretation	Either party may apply the architect for an interpretation.
2.2.9.		Respond in writing in a "reasonable time" on receipt of claims or disputes between owner and contractor	All disputes between the owner & contractor go initially to the architect re: execution/progress of the work or interpretation.
2.2.10.		Make interpretations impartially & consistent with the contract documents.	The architect carries no liability for decisions made for interpretations if made in good faith & consistent with the contract documents.
2.2.11.		Make decisions related to artistic effect.	Architect's decision on artistic effect is final if consistent with contract documents
2.2.12.		Deliver decisions in writing within 10 days	Most claims subject to arbitration (not 2.2.11.).
2.2.13.		Reject work not conforming to contract documents. Special inspection and testing powers (7.7.2)	Architects decision here implies no responsibility to contractors or subcontractors or their agents or employees
2.2.14.		Review/approve shop drawings, product data, samples, etc. with reasonable promptness.	Approval of item *not* indicative of approval of whole assembly.
2.2.15.		Prepare Change Orders (see art. 12) and make minor changes (i2.4.1)	
2.2.16.		Conduct inspections for substantial completion and final completion. Receive and forward warranties, etc. to owner. Issue final certificate for payment (9.9)	The architect has no authority to pass comment upon the legal sufficiency of warranties
2.2.17.		The architect may have a representative on site if the owner *agrees*.	Duties and responsibilities of the project representative set out in the contract documents
2.2.18.		Architect's duties etc. are specified in the contract documents	The architect's duties/tasks are *not* extended beyond those agreed unless written consent by architect, owner and contractor.
2.2.19.			Appointment of a new architect only if contractor has no reasonable objection. Disputes subject to arbitration.

(continued)

The Articles

ARTICLE/ PARAGRAPH/ SUBPARA	CONTENT	ARCHITECT'S POWERS/ DUTIES	COMMENT
ARTICLE 3	**OWNER**		
3.2.	Information and Services Required of the Owner	Ensure owner is aware of all *his* obligations 3.2.1, 3.2.2, etc.	
3.2.6.		Forward all communications from the owner to the contractor.	
3.3.1.	Owner's Right to Stop Work		The architect is *not* empowered to stop the work unless specifically authorized by the contract docs.
3.4.1.	Owner's Right to Carry Out the Work	Approve owner's intention to carry out work and amounts charged to the contractor. Issue change order deducting cost from future payments.	The architect is entitled to extra compensation for additional services rendered in respect of this provision
ARTICLE 4	**CONTRACTOR**		
4.1.1.	Definition		The contractor's representative should be identified to the owner and architect.
4.2.1.	Review of Contract Documents		Any error spotted by the contractor in the contract documents reported *at once* to avoid any responsibility.
4.4.1.	Labor and Materials		The contractor must provide adequate labor, materials etc., to do the work or 2.2.13 (reject), 3.3.1 (owner stop work), 3.4.1 (owner do work), 7.6.1, 14.2.1 (termination).
4.5.1.	Warranty	May require evidence of quality and conformity of materials/equipment from the contractor.	
4.7.3.	Permits, Fees, & Notices	If necessary, the architect should issue modifications to bring contract documents in conformity with all applicable laws and codes	If the contract documents do not conform to laws etc., contractor report to architect promptly in writing.
4.7.4.			If contractor does work knowingly contrary to laws etc. without notifying architect, he assumes full responsibility.
4.8.2.	Allowances	Prepare and issue change order if required.	All allowances included on the contract sum. If more or less, the contract sum adjusted by change order.
4.10.1	Progress Schedule	Take receipt of estimated progress schedule and send copy to owner.	Contractor to send progress schedule to architect/owner *as soon as* contract is awarded, for information *not* approval.
4.11.1.	Documents and Samples at the Site	Check all documents returned on completion.	One copy of all drawings, specs, change orders etc. kept on site with shop drawings etc. for the owner and delivered to architect when work is completed.
4.12.6	Shop Drawings, Product Data, & Samples	Approve in writing specific deviations if written request by contractor considered acceptable	The architect's approval does not prevent the contractor for being responsible for nonconformity with the contract documents unless architect gives written approval.
4.12.8.		Approval of submittals (2.2.14) before work started.	No work requiring shop drawing etc. started until architect's approval.
4.16.1	Communications	Forward all communications from the contractor to the owner.	
4.17.1.	Royalties and Patents		If the contractor believes a patent may be infringed, he must inform architect or take responsibility. Architect may share liability for infringement as owner's agent.
4.18.1.	Indemnification	Architect indemnified against actions based on the negligent performance of the work.	The contractor indemnifies the owner and architect to the extent of the law. (State statutes vary on this.)
4.18.2.		This indemnification cannot be limited to a financial amount	
4.18.3.			Contractor's obligations do not extend to architect's responsibilities i.e., preparation of drawings, change orders, etc.

(continued)

ARTICLE/ PARAGRAPH/ SUBPARA	CONTENT	ARCHITECT'S POWERS/ DUTIES	COMMENT
ARTICLE 5	SUBCONTRACTORS		
5.2.1.	Award of Subcontracts and Other Contracts for Portions of the Work	Promptly reply if any reasonable objection by architect or owner.	Contractor informs owner & architect "as soon as practicable" of proposed subcontractors. No reply means no objection.
5.2.2.			The contractor may not work with anyone reasonably objected to by the architect or owner, or be required to work with anyone he, the contractor, reasonably objects to.
5.2.3.		Prepare change order if the contract sum is increased or decreased as a result of a substitution.	If prompt substitution is made by the contractor after architect or owner objection, the contract sum may be adjusted.
5.2.4.		Reasonable objection to subcontractor, substitution can be made.	No substitution if reasonable objection by owner or architect.
5.3.1.	Subcontractual Relations		Subcontractors are required to be bound by same contract conditions and obligations as the contractor.
ARTICLE 6	WORK BY OWNER OR BY SEPARATE CONTRACTORS		
6.1.1.	Owner's Right to Perform Work & Award Separate Contracts		Separate contracts involve extra work by the architect, and merit additional compensation.
6.1.3.		The architect could be the coordinator of the owner's workforce (for additional compensation).	Owner coordinates own forces and separate contractors with the work of the contractor.
6.2.2.	Mutual Responsibility		If the contractor can't work due to owner/separate contractor fault, he must report *promptly* to the architect or acceptance is deemed.
6.3.1.	Owner's Right to Clean Up	Architect determines who pays the cost of cleaning up	The owner may clean up the site if a dispute arises between the contractor and subcontractors.
ARTICLE 7	MISCELLANEOUS PROVISIONS		
7.1.1.	Governing Law		The law of the place where the work is located governs the project.
7.5.1.	Performance Bond & Labor & Material Payment Bond		The owner has the right to ask for bonds. Advice on this is outside the architect's duties.
7.6.1.	Rights and Remedies		No right or duty under the contract is waived by the action or failure to act by the owner, contractor, or architect.
7.7.2.	Tests	May observe inspections, testing, or approvals.	If tests/inspections are required by law, the contractor must give the architect "timely notice."
7.7.2.		If special tests are deemed necessary outside 7.7.1. get written authority from owner and instruct contractor to order inspection. If work passes, issue change order for extra work.	If work fails inspection, the contractor pays for tests, including architect's additional compensation. If work passes, owner pays.
7.7.3.			Required certificates of inspection etc. "promptly" delivered to architect by the contractor.
7.7.4.		Observe inspections etc. where required "promptly" and where possible at source of supply.	
7.9.1.	Arbitration		The architect is *not* included in any arbitration agreement between the owner and contractor unless by written consent.
7.9.2.			A copy of any arbitration demands should be filed with the architect.

(continued)

The Articles

ARTICLE/ PARAGRAPH/ SUBPARA	CONTENT	ARCHITECT'S POWERS/ DUTIES	COMMENT
ARTICLE 8	**TIME**		
8.1.1.	Definitions	If required, issue change order to adjust substantial completion date.	The contract time can only be changed by change order.
8.1.2.			The work is started by a notice to proceed, or date in owner/contractor agreement or other date agreed upon.
8.1.3.		Certify the date of substantial completion (or designated portion).	
8.3.1.	Delays & Extensions of Time	Determine justifiable delay and issue change order for "reasonable time."	The contractor may get an extension of the contract time for certain reasons, including ones that the architect determines are just.
8.3.2.			Claims for extensions made in writing to architect within 20 days of start of delay or waived. Estimates of probable effect of delay required.
8.3.3.		If required, the architect has 15 days to render an interpretation (2.2.8) unless otherwise specified.	
ARTICLE 9	**PAYMENTS & COMPLETION**		
9.1.1.	Contract Sum	If required, issue change orders to adjust contract sum (12.1.1).	The contract sum may only be changed by change order.
9.2.1.	Schedule of Values	If appropriate, require evidence of accuracy. Object to schedule if dissatisfied.	Before first application for payment, contractor to submit to architect a schedule of values for portions of the work with such data as the architect may require.
9.3.1.	Applications for Payments	If appropriate according to owner/architect, require supporting data.	Contractor submits application for payment to architect at least 10 days before agreed date for progress payment. Supporting data required by architect (vouchers, release of liens, etc.) should be included.
9.3.2.		Check data for off-site or nonincorporated materials and equipment before issuing certificate.	Payment for materials and equipment not incorporated in the work or off-site is permissable, but bills of sale, insurance, etc. may be required.
9.4.1.	Certificates for Payment	Issue certificate within 7 days of application to owner (copy to contractor). Determine amount or notify contractor for withholding reasons (9.6.1).	
9.4.2.			Certificate indicates only approval to *the best of the architect's knowledge*. *Not* exhaustive inspection or knowledge of how the contractor has spent previous money.
9.5.2.	Progress Payments		*No architectural duty* in disputes between contractor/subcontractor etc. unless owner/contractor agreement breached.
9.5.3.		*On request and at his discretion,* the architect may furnish subcontractors with data on amounts applied for by the contractor and the architect's action on the work done by the subcontractor.	
9.5.4.			*No obligation* by the architect/owner to pay or see to payments to subcontractors (unless required by law).
9.6.1.	Payments Withheld	The architect may decline to certify payment and refuse to issue certificates in whole or in part. If so, the contractor should be notified and if no revised agreement can be reached, the architect must promptly certify the amount he considers acceptable. The architect can also nullify previous certificates to protect owner from loss (defective work unremedied etc.).	
9.7.1.	Failure of Payment	The architect should issue a change order to adjust the contract sum when the work is restarted.	If no certificate issued through no fault of the contractor, he may give 7 days notice (written) and then stop work until paid. Additional pay due to the shutdown authorized by change order.

ARTICLE/ PARAGRAPH/ SUBPARA	CONTENT	ARCHITECT'S POWERS/ DUTIES	COMMENT
9.8.1.	Substantial Completion	Inspect upon notification. If satisfied with work and punch list, certify, and state date by which work should be completed. Submit certificate of substantial completion to owner and contractor for written acceptance.	When substantial completion (part or whole) is reached, the contractor will submit a list of items to be completed or corrected to the architect (punch list).
9.9.1.	Final Completion & Final Payment	Upon written notice, the architect inspects for final completion. If satisfied, promptly issue final certificate for payment of the balance.	
9.9.2.			The contractor will submit to the architect various data (consent of sureties, affidavits etc.) before final payment and retainage released.
9.9.3.		Architect may confirm delay and certify for payment.	If final completion is delayed through no fault of the contractor, or issuance of change orders, payment may be made for completed work.
ARTICLE 10	PROTECTION OF PERSONS & PROPERTY		
10.1.1.	Safety Precautions & Programs		Safety is the *exclusive* responsibility of the contractor. The architect should not detail any safety measures.
10.2.6	Safety of Persons & Property		If the contractor's representative for safety measures is other than the superintendent, he must notify the architect and owner in writing.
10.3.1.	Emergencies	The architect may determine additional money and time for emergency work and issue a change order to that effect.	
ARTICLE 11	INSURANCE		
11.1.4.	Contractor's Liability Insurance		Certificates of insurance from the contractor should be sent to the owner (4.16.1) before work begins.
11.3.1.	Property Insurance		The owner maintains all risk property insurance. The architect's liability insurance generally prohibits advice on insurance and bonds.
11.3.5.		Issue change orders to reduce the contract sum (if required) for the cost of further insurance cover requested by the contractor.	The contractor may request further insurance cover than 11.3.1 and 11.3.2. The owner will act, and charge the cost to the contractor.
11.3.6.			The AIA warns that some building risk policies may override waiver of rights. Owners should be made aware of this potential problem so they may seek legal and insurance advice.
11.3.7.		Issue change orders for the replacement of damaged work.	The owner may give bond after an insured loss for the proper performance of his duties.
ARTICLE 12	CHANGES IN THE WORK		
12.1.1.	Change Orders	Send *all* change orders to the owner for signature before issuance.	Change orders must be signed by both owner and architect to effect changes in the contract sum or contract time.
12.1.3.		Check the contractor's claims and recommend to the owner acceptance, rejection, or negotiation. If accept, prepare change order and forward notice to proceed from owner.	The cost resulting from the change in the work may be determined by: • lump sum • unit prices • fixed or percentage fee cost agreed upon plus • method expressed in 12.1.4.
12.1.4.		Determine "reasonable" cost and overhead profit and ask the contractor for proof (an itemized account for expenditure) before including in a change order.	If none of the formulas in 12.1.3 are acceptable, upon signed order from the owner, the contractor will proceed with the work and the architect will determine a reasonable cost and profit sum.
12.1.5.	Architect's Powers and Duties	If unit prices are agreed upon, but subsequent changes render the originally agreed prices inequitable, the architect may adjust the prices in the change order.	
12.2.1.	Concealed Conditions	Issue change orders to amend the contract sum in respect of concealed conditions.	If concealed conditions as described are found, the contract sum may be "equitably" adjusted by change order if conditions are reported within 20 days of discovery.

(continued)

The Articles

ARTICLE/ PARAGRAPH/ SUBPARA	CONTENT	ARCHITECT'S POWERS/ DUTIES	COMMENT
12.3.1.	Claims for Additional Costs	Determine additional costs if there is no agreement between owner and contractor, and issue change order.	Claims for additional cost by the contractor must be made to the architect within 20 days of the event in writing.
12.4.1.	Minor Changes in the Work	Make minor changes without consulting the owner only on matters not affecting the contract sum or contract time (in writing) not inconsistent with the intent of the contract documents.	Minor changes made by the architect bind the owner and contractor. The written orders shall be carried out "promptly."
ARTICLE 13	UNCOVERING AND CORRECTION OF WORK		
13.1.1.	Uncovering of Work	Write to the contractor if appropriate to request uncovering of work for observation.	Uncovering will be at the contractor's expense if the work was covered contrary to the architect's request or the contract documents (supplementary conditions).
13.1.2.		Request other work not specifically mentioned in the contract documents to be uncovered. If the work conforms, issue a change order for the additional work and cost.	If the work requested for uncovering conforms, the owner pays. If it does not, the contractor pays.
13.2.1.	Correction of Work		Work rejected by the architect must be remedied "promptly" by the contractor. Architect's additional expenses also charged to contractor.
13.2.5.		Fix "reasonable" time in which to correct defective or nonconforming work in writing. Issue change order if sales of contractor's equipment do not cover costs contractor should have borne if owner forced to sell equipment, and balance charged to contractor.	If the contractor does not remedy the defective work in the reasonable time, the owner can remove and store materials and equipment. If within 10 days the contractor has not paid for removal or storage, the owner, after 10 days additional written notice, can sell the materials and equipment in question.
13.3.1.	Acceptance of Defective or Nonconforming Work	Issue change order reducing the contract sum if required to accept defective or nonconforming work by the owner's request.	
ARTICLE 14	TERMINATION OF THE CONTRACT		
14.1.1	Termination by the Contractor		If the work is stopped for any of the stated reasons for 30 days, the contractor may, after 7 days notice in writing to the architect and owner, terminate and recover costs from the owner.
14.2.1.	Termination by the Owner	Certify "sufficient cause" exists for the owner to terminate the contract.	If the contractor breaches the stated conditions, the architect may certify that sufficient cause exists for the owner, after 7 days written notice to the contractor, to terminate the contract and finish the work. The AIA warns that this may be void under the Federal Bankruptcy Act of 1979. Termination should be carried out with the assistance of legal counsel.
14.2.2.		Certify the amounts to either the owner or contractor to settle the financial relationship between them.	If the unpaid balance exceeds the cost of finishing the work (including architect's additional fees), the contractor is paid the excess. In reverse circumstances, the contractor must pay the difference. This is certified by the architect upon application (9.4).

During the development of each scheme, it will be necessary to establish on what basis the project will be constructed. This may have been discussed with the owner much earlier in the process, in relation to other important factors, such as time or finance available, type of project, or site characteristics.

Once such variables have been assessed, certain alternative construction procedures can be considered. These might include:

- Single or separate contract systems
- Negotiated or competitively bid contracts
- Types of building contracts (see page 75).

Single or Separate Contract Systems

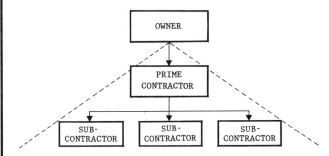

Single Contracts

In most construction projects, a *prime contractor* is responsible for the full extent of the construction involved. If the work requires more labor or skill than the prime contractor can supply, *subcontractors* (and even *sub-subcontractors*) will be hired, but will remain the responsibility of the prime contractor in matters of liability to the owner, payment, etc. (see page 97).

Separate Contracts

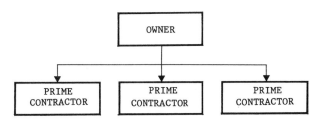

In some instances (e.g., certain state work and some larger projects), contractors will be selected for specific and distinct divisions of the work (electrical, mechanical, etc.). There is no prime contractor as such because all contractors will have an equal relationship with the owner. This system has the advantage of reducing the prime contractor's extra charge for administration of the subcontractors and reduces the expense of double insurance. However, it may also complicate the relationship between the contractors involved. Since no hierarchy of responsibility exists between them, management and supervision of the project have to be coordinated carefully to ensure smooth and uninterrupted transfer between the individual work forces. Problems have been known to occur in matters of delay, clearing up, etc., and the role of the *construction manager* (see page 28) has recently become useful in this contracting system as a coordinator and supervisor of the various work forces. Alternatively, the architect could be employed by the owner to undertake this task as part of additional services.

Negotiated or Competitively Bid Contracts

The selection of the contractor can be undertaken in two ways, depending upon the character of the project:

Negotiated Contracts

The owner can select a contractor directly based upon the latter's reputation, recommendation, etc., and then negotiate the terms of agreement and form of payment. This may be appropriate:

- Where the contractor possesses skill or experience relevant to the project
- Where quality and not economy is a major determinant
- Where an early completion of the project is desired
- Where details of the final scheme are incomplete.

Using the direct selection approach, the owner need not wait until the normal selection phase to begin construction. This means that completed drawings and specifications are not necessary for work to start, and that the contractor's skill and expertise may be brought into the design process.

However, there is no competition in this form of selection, making it potentially unsuitable where a low overall price is sought. Usually, a *cost plus fee* contract (see page 75) is used in conjunction with negotiated contracts.

Competitively Bid Contracts

In order to obtain the lowest possible price for the work, completed sets of contract documents are sent to a number of contractors who bid against each other. Usually, the lowest bidder is awarded the contract (see page 89).

Public agencies often require this method of contracting, which is best suited to projects of a straightforward, traditional nature, where no unforeseen problems are likely.

Contractor Selection

It is widely used in the construction industry, and necessitates complete design documentation to enable accurate bidding.

In the last few years, certain alternatives to the basic contracting methods have emerged. These include:

- Fast tracking
- Design/build work
- Turnkey contracts.

Fast Tracking

As previously stated, if the contractor is selected on a negotiated basis, it is possible to begin construction work before the completion of the design phase. This method of overlapping the design and construction work is known as "fast tracking".

Design/Build Work

In the traditional model of building, the phases of design and construction are separate and usually undertaken by different specialists, and therefore different firms. Some companies now provide a package combining all the functions of the building process, often in large specialist-type projects, allowing the owner to contract with only one party to provide the complete building. This has the advantage of allowing fast tracking, and combining the skills of the designers and the constructors. It does, however, deprive the owner of an expert agent to look out for his/her best interests throughout the project.

Turnkey Contracts

This kind of contract usually relates to projects where a developer proposes and constructs an entire development (including the purchase of the site) and hands it over to the owner ready for immediate occupancy when complete. It has been used in dealings with local housing authorities.

In order to facilitate the process of contractor selection, the AIA has developed AIA Document G611a, Owner's Instructions Regarding Bidding Documents and Procedures. This form may be filled out by the owner in advance of the selection stage, and can help to clarify the requirements and preferences for the architect so that appropriate action can be advised.

References

Sweet, pp. 256–294, 246.
Walker, pp. 148–151.
AIA 16.

If the contractor is going to be selected by competitive means, certain procedures can be implemented to facilitate the task.

Selection of Bidders

The initial process of identifying possible bidders may be:

- Open
- Selective.

Open Bidding

Where the maximum number of bidders is considered desirable (usually in public work), an *advertisement to bid* will be published in trade or governmental publications or professional journals, inviting any interested contractors to participate in the process.

Selective Bidding

If a limited number of bidders is preferred (the AIA suggests that 6 bidders are adequate in most cases), an *invitation to bid* will be sent to a number of contractors. These will be singled out by reputation, recommendation, previous contact with either the owner or the architect, etc.

Contractor Qualification

Prospective bidders should be chosen for their ability to successfully undertake the project, and it may be necessary to establish their suitability before bidding documents are issued. In some cases, the contractor's reputation or relationship with the owner will be sufficient, but AIA Document A305, Contractor's Qualification Statement, may help to outline contractors' suitability.

The document, when completed, provides full details of the contractor's business record, and enables the owner and the architect to gain a clear impression of such details as:

- History of the business
- Organization and scope of operations
- Past record of construction work (type of work, range of experience, etc.)
- Trade and bank references
- Bonding company
- Details of assets and liabilities.

The qualification statement can be used as a pre-qualification stage in the open bidding process to eliminate unsuitable bidders and cut down the administration involved in high numbers.

Once the bidders have been identified and contacted to ensure their interest, a package of information concerning the proposed project is issued. The package includes:

- The invitation/advertisement to bid
- Drawings and specifications (see pages 66–67)
- The bid form
- Notice to Bidders
- Instructions to Bidders
- Proposed contract documents
- Bid security details (if required).

Drawings and Specifications

These documents, which should be as complete and unambiguous as possible to allow the contractors to bid accurately, are sent free of charge to the bidders. The number of sets necessary for each bidder varies, although the AIA suggests:

Small projects (below $300,000)	2 sets
Large projects ($300,000–$1,000,000)	3 sets
Larger projects (above $1,000,000)	3 + sets

In some cases (where, for example, time or complexity are major factors of importance), more sets may be required to expedite the bidding process. The architect can require additional payment for the extra work necessary to accomplish this. Similarly, if any of the bidders asks for extra copies, they may be provided at their expense.

To ensure return of the bidding documents by unsuccessful bidders, a deposit is usually required which is returned upon receipt of outstanding documents.

Notice to Bidders

This may be included in the bidding documents, and informs prospective bidders of their opportunity to bid, and the conditions and requirements involved.

Instructions to Bidders

AIA Document A701, Instructions to Bidders, provides all relevant information concerning the detailed requirements involved in the bidding process, including:

- Definitions
- Bidding documents
- Consideration of bids
- Owner–contractor agreement
- Supplementary instructions
- Bidder's representations
- Bidding procedures
- Post-bid information
- Performance/labor and material payment bonds (see page 95).

Bidding 1

Contract Documents

All documents intended for use in the proposed project should be sent to each bidder for examination, including the conditions (e.g., AIA Document A201) and any other applicable addenda or supplementary conditions.

Bid Form

This form, which should be sent to all bidders, contains all relevant information concerning the project. Each bidder will then return the document complete with the price of the work, or *base bid*, and any other figures which may be appropriate (e.g., alternate bids, substitutions, etc.).

Bid Security

In order to ensure each bidder's commitment to their base bid, some form of security may be required by the owner, which should be submitted along with the returned bid form. The security might take the form of cash, a certified check, or a *bid bond* (AIA Document A310; see pages 93 and 95). The bond could be expressed either in a lump sum or as a percentage of the base bid, although the former is usually preferred by bidders, as it does not reveal their bid before opening. The bond ensures that, in the event of the successful bidder refusing to undertake the work for the bid specified, the whole or part of the security may be forfeit. The amount of the penalty is usually determined as the difference between the selected bid and the next lowest.

Variations

Where possible, documentation necessary for accurate bidding should be comprehensive and unambiguous. In some instances however, it may be necessary to provide some alternatives in the bidding process if requirements cannot be fully determined. Two mechanisms which allow this are:

- Alternates
- Unit prices.

Alternates

An alternate bid may be required or accepted for a specified section of the work, and should be included in the calculation of the base bid. This procedure can be useful in helping to keep costs within a certain budget, but should be used sparingly and not employed to give one bidder preference over the others.

Unit Prices

Unit prices provide a means of measurement which can be included in the bid, indicating a price per unit for materials and/or services. It is useful in giving an idea of price calculation for unknown quantitites or variable factors and, again, should be restricted in use if the overall budgetary figure needs to be controlled.

References

Sweet, pp. 256–294.
AIA 16.

In implementing the AIA procedures of contractor selection by requesting bids, certain rules have been developed which should be adhered to by all parties concerned. These are outlined in AIA Document A701, Instructions to Bidders.

The procedural format, following the mailing of necessary bidding documents includes:

- Modification of bidding documents
- Submission of bids
- Bid opening
- Selection
- Announcement
- Contract award.

Modification of Bidding Documents

Certain queries or adjustments to the documents might be necessary or requested prior to the closing date for submission. These are usually in the form of:

1. Interpretations
2. Substitutions.

1. Interpretations

If any of the bidders should discover errors or ambiguities in the documentation, they must inform the architect in writing, at least 7 days prior to the submission date. Any changes or addenda will then be issued by the architect to all bidders.

2. Substitutions

Should any of the bidders wish to substitute materials or services otherwise specified in the bidding documents, the architect must receive a request for approval in writing, at least 10 days prior to the submission date. If the architect decides that the substitution is acceptable, all parties will be notified by addendum, although no addenda can be made within 4 days of the final receipt date, except a notice cancelling or postponing the request for bids.

Submission of Bids

Bids must be delivered in writing, contained in sealed, opaque envelopes prior to the time and date specified in the advertisement/invitation to bid. Oral bids are not acceptable. Any bids received after the specified time should be returned unopened.

Bid Opening

If the bids are opened in public, they are often read aloud, whereas if opened in private, the bidding information may be sent to all bidders at the owner's discretion. The owner need not accept any of the bids if they appear too high, and may reject any bid not in conformance with the stated requirements. The bidding documents do provide, however, that if a contractor is chosen, it will be on the basis of the *lowest responsible bid*. The decision is usually reached within 10 days of the bid opening.

In publically bid work, the owner is often constrained by law to accept the lowest responsible bidder, and may be held criminally liable if the selection does not conform to these requirements (i.e., the lowest monetary bid, coupled with the owner's satisfaction that the contractor can successfully undertake the work). In privately bid work, the commitment is not as clear, although the rules of bidding should be adhered to. Granting of the contract to any other than the lowest bidder should only be made with very good reason to prevent suspicion of favoritism, and ill-feeling among the contractors.

Selection

At any time prior to the bid opening, all bidders may withdraw or modify their bids. However, once the bids are opened, the bidders cannot make changes or withdraw from the process for a period stipulated in the bidding documents (e.g., 30 days). Once selected, the successful bidder must undertake the work for the agreed price, or risk forfeiture of the bid bond (if any). Exceptions to this are sometimes made if the bidder can prove substantial error in the bid calculation, in which case, withdrawal might be appropriate, with award of the contract to the next lowest bidder. Alternatively, the contract may be rebid. Defaulting bidders should be disqualified from any further bidding on the same project, and no bid correction should be permitted, except for minor clerical errors and alterations.

Announcement

When a contractor has been selected (usually within 10 days of bid opening), all bidding parties should be informed of the decision. The unsuccessful bidders are often given a list of the bid figures, and the bid deposits are returned once the documentation is received. The successful bidder should be informed of the decision in a way which does not form a legally binding agreement prior to the signing of the contract documents. Usually, the bids of the next two or three lowest bidders will be retained for a period as a contingency measure.

At this stage, each party to the proposed building contract may provide further information and/or assurances to the other:

The owner will, upon request, prove to the contractor that sufficient financial arrangements have been made to undertake the project.

Bidding 2

The contractor, within 7 days of the contract award, should furnish:

- Details of the amount of work to be undertaken by the contractor's forces
- Names of proposed suppliers of materials and equipment
- A list of intended subcontractors for the architect's approval (see page 97).

The contractor may also be asked for:

- A contractor's Qualification Statement (if appropriate and if not completed prior to selection)
- Proof of the responsibility and reliability of the work force
- Bonds, in accordance with the owner's requirements as expressed in the Instructions to Bidders.

When these preliminary matters have been dealt with, and the contract documents are prepared, both parties will be ready to enter into the contractual agreement.

References

Walker, pp. 148–151.
AIA 16.

The contract documents comprise:

- The owner–contractor agreement
- The conditions of the contract (including any supplementary details or other conditions)
- The drawings
- The specifications
- Any addenda previously issued, or modifications (i.e., written amendments to the contract, signed by both parties, e.g., change orders, written interpretations, or minor changes)
- Related documents and agreements
- Performance bond and labor and material payment bond
- Owner's insurance and contractor's insurance.

When the documents are ready, they should be sent to the parties for signing with a covering letter.

Notice to Proceed

This is a written authorization from the owner to the contractor establishing a date of commencement and completion of the building work. The Notice to Proceed is used if the work is started *after* (not before) the date of the signing of the contract.

Letter of Intent

Should construction need to be started *before* the contract documents have been signed (e.g., where time is of the essence), a letter of intent may be sent by the owner, giving the contractor authority to proceed. If used, the letter should be carefully drafted to avoid any conflict with the actual contract documents, and legal assistance should be sought. The letter should emphasize that no subcontracts should be effected, nor should any materials be ordered other than those relevant to the specific work permitted. Insurance should be carefully considered if a letter of intent is used, and it should be made very clear that the letter will cease to have effect upon the signing of the actual contract.

Once the contractual relationship is established, certain obligations must be met by both parties, including:

- Owner capability (see page 89)
- Contractor's work schedule
- List of subcontractors
- Schedule of values
- Certificate of insurance
- Permits.

Contractor's Work Schedule

As soon as the conract has been awarded, the contractor should provide for the architect's information an estimated schedule of progress. This is usually in the form of:

- A bar chart
- Critical path method.

Bar Chart

A bar chart indicates the work, divided by trades or operations, against which a time scale can be set. The progress of the work can be plotted between the two.

Critical Path Method

Critical path analysis is a project planning device which aims to optimize time and operations on site. The system divides various activities which are sequenced in terms of their interrelationship. When the time factor is added, a path may be plotted which reveals the most efficient operational procedures which should be followed. The schedule can be monitored by regular assessment of actual achievement on site. This enables continued prioritizing and adjustment throughout the period, to enable maximum efficiency in allotting time for the various stages of the project.

PERT Project Evaluation Review Technique)

This is a method of scheduling which establishes, in chart form, activities and operations anticipated in the project layout which can introduce a cost element into the programming. PERT has not been commonly adopted in the construction industry.

List of Subcontractors (see page 97)

Schedule of Values

Prior to the first application for payment, the contractor must submit a schedule of values to the architect, together with any data supporting its accuracy that the architect may require. This then forms the basis for reviewing future applications for payment, and should indicate the sections of the contract sum provided for the various parts of the work.

Certificate of Insurance

The contractor should file with the owner (or architect) certificates of insurance before starting work (AIA Document A201, Article 11.1.4; see page 94).

Contract Procedures

Permits

Under the AIA general conditions, the contractor is responsible for obtaining the building permit and certain other governmental requirements, e.g., licenses.

Reference

AIA 17, 18.

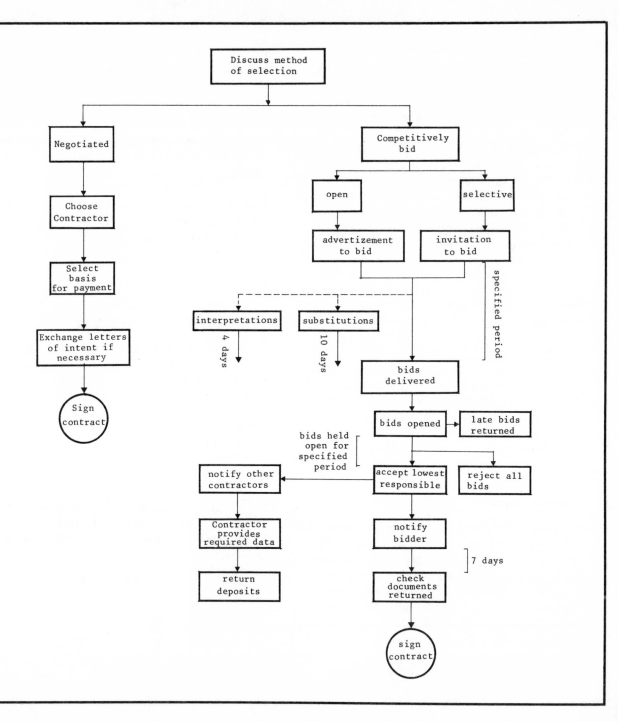

Bid Bond

THE AMERICAN INSTITUTE OF ARCHITECTS

AIA Document A310

Bid Bond

KNOW ALL MEN BY THESE PRESENTS, that we

PHILIP DA TRENCHYN INC. WITTS END, WISCONSIN
as Principal, hereinafter called the Principal, and
(Here insert full name and address or legal title of Contractor)

ACME GUARANTEE CO. MILWAUKEE, WISCONSIN
(Here insert full name and address or legal title of Surety)
a corporation duly organized under the laws of the State of WISCONSIN
as Surety, hereinafter called the Surety, are held and firmly bound unto

ELLEN I. WATER, P.O. BOX 314, HOLDEMAT BAY, WISCONSIN
as Obligee, hereinafter called the Obligee, in the sum of
(Here insert full name and address or legal title of Owner)

FIFTEEN THOUSAND————————————— Dollars ($ 15,000.00),
for the payment of which sum well and to be made, the said Principal and the said Surety, bind ourselves, our heirs, executors, administrators, successors and assigns, jointly and severally, firmly by these presents.

WHEREAS, the Principal has submitted a bid for
(Here insert full name, address and description of project)
WATER RESIDENCE 1. LAKESIDE
HOLDEMAT BAY, WISCONSIN

NOW, THEREFORE, if the Obligee shall accept the bid of the Principal and the Principal shall enter into a Contract with the Obligee in accordance with the terms of such bid, and give such bond or bonds as may be specified in the bidding or Contract Documents with good and sufficient surety for the faithful performance of such Contract and for the prompt payment of labor and material furnished in the prosecution thereof, or in the event of the failure of the Principal to enter such Contract and give such bond or bonds, if the Principal shall pay to the Obligee the difference not to exceed the penalty hereof between the amount specified in said bid and such larger amount for which the Obligee may in good faith contract with another party to perform the Work covered by said bid, then this obligation shall be null and void, otherwise to remain in full force and effect.

Signed and sealed this   FIFTH   day of MAY   1981

P da Trenchyn   (Seal)
(Principal)
PRESIDENT
(Title)

D.A.Smith
(Witness)

Jno W James
(Surety)   (Seal)
PRESIDENT.
(Title)

J.C. Brown
(Witness)

1

93

Certificate of Insurance

CERTIFICATE OF INSURANCE
AIA DOCUMENT G705

This certificate is issued as a matter of information only and confers no rights upon the addressee. It does not amend, extend or alter the coverage afforded by the policies listed below.

Name and Address of Insured

PHILIP DA TRENCHYN, WITTS END, WISCONSIN

Covering (Project Name and Location)

WATER RESIDENCE, 1 LAKESIDE, HOLDEMAT BAY, WI

Addressee: (Owner)
ELLEN I. WATER
P.O. BOX 314
HOLDEMAT BAY, WISCONSIN

COMPANIES AFFORDING COVERAGE

A	ACME GUARANTEE CO.
B	PRUDENT INSURANCE & CASUALTY
C	
D	
E	
F	

This is to certify that the following described policies, subject to their terms, conditions and exclusions, have been issued to the above named insured and are in force at this time.

TYPE OF INSURANCE	CO. CODE	POLICY NUMBER	EXPIRATION DATE	LIMITS OF LIABILITY IN THOUSANDS	EACH OCCURRENCE	AGGREGATE
1. (a) Workers' Compensation		AC. 1473	12/31/81	Statutory		Each Accident
(b) Employer's Liability		AC. 2221	12/31/81			
2. Comprehensive General Liability including:		PR 123.11	12/31/81			
☒ Premises - Operations				Bodily Injury	$ 50,000	$ 300,000
☐ Independent Contractors				Property Damage	$ 100,000	$ 500,000
☐ Products and Completed Operations						
☒ Broad Form Property Damage				Bodily Injury and Property Damage Combined	$	$
☐ Contractual Liability						
☐ Explosion and Collapse Hazard						
☐ Underground Hazard				*Applies to Products and Completed Operations Hazard		
☐ Personal Injury with Employment Exclusion Deleted					$	$ (Personal Injury)
3. Comprehensive Automobile Liability				Bodily Injury (Each Person)	$	
☐ Owned	NONE	N/A	N/A	Bodily Injury (Each Accident)	$	
☐ Hired				Property Damage	$	
☐ Non-Owned				Bodily Injury and Property Damage Combined	$	
4. Excess Liability				Bodily Injury and Property Damage Combined	$	$
☐ Umbrella Form	NONE	N/A	N/A			
☐ Other than Umbrella						
5. Other (Specify)	NONE	N/A	N/A			

1. Products and Completed Operations coverage will be maintained for a minimum period of ☒ 1 ☐ 2 year(s) after final payment.

2. Has each of the above listed policies been endorsed to reflect the company's obligation to notify the addressee in the event of cancellation or non-renewal? ☒ Yes ☐ No

CERTIFICATION

I hereby certify that I am an authorized representative of each of the insurance companies listed above, and that the coverages afforded under the policies listed above will not be cancelled or allowed to expire unless thirty (30) days written notice has been given to the addressee of this certificate.

ACME GUARANTEE CO.
Name of Issuing Agency

MILWAUKEE, WISCONSIN
Address

_____
Signature of Authorized Representative

4/15/81
Date of Issue

Of the various measures often taken in the construction industry to minimize risk and potential loss, surety bonds are a common precaution. A surety bond is basically an assurance by one party which provides that specified obligations of another will be met, despite unforeseen or undesirable events. In reality, the cost of bonds, although technically borne by the contractor, is transferred to the owner by inclusion in the bid.

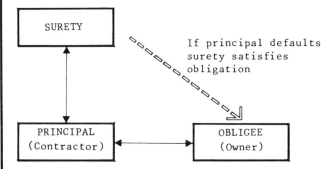

Types of Bond

Three bonds used frequently in the construction industry are:

1. Bid bond
2. Performance bond
3. Labor and material payment bond.

1. Bid Bond (see page 93)

In order to ensure that the selected bidder signs the contract and fulfills other preliminary requirements, a bid bond may be requested. This should cover not less than 10% of the bid amount, and would be used to pay the owner the difference between the two lowest bids if the successful bidder decides to back out. The penalty for this cannot exceed the bond amount, which should be expressed as a specific sum, not a percentage of the bid.

2. Performance Bond

A performance bond ensures that, in the event of the contractor failing to fulfill the contractual requirements, the surety posting the bond amount guarantees that the owner will be financially protected up to the amount expressed as the bond penalty. This bond is usually used in conjunction with:

3. The Labor and Material Payment Bond

This bond ensures that all bills for labor and materials will not revert to the owner in the event of nonpayment by the contractor.

Combination bonds are considered inadvisable by the AIA as they can cause legal complications in the event of a claim. The AIA recommends the two-bond system as a preferable procedure. State laws should be checked regarding the use of bonds, as statutory requirements vary with regard to bond provisions.

If claims are made against bonds during a construction project, AIA Document B141, Owner–Architect Agreement, provides for additional payment to the architect for the work involved in making the necessary arrangements for the continuation of the project.

Other Bonds

Other forms of bond sometimes used include:

- License or permit bond
- Lien and no-lien bond
- Maintenance bond
- Release of retained percentage bond
- Statutory bond (check each state for requirements)
- Subcontract bond
- Termite bond.

AIA Bonds

Although there are no standardized requirements for bonds compatible with all state laws and owner preference, the AIA produces certain forms which are helpful in many cases. These include:

- A310, Bid bond (see page 93)
- A311, Performance Bond (see page 96) and Labor and Material Payment Bond
- A311/CM, Performance Bond and Labor and Material Payment Bond, Construction Management Edition.

In certain states, variations of the basic forms have been developed to comply with individual state laws for use in public and private construction projects.

In all matters relating to bonds and insurance the owner should seek expert advice. The architect should not attempt to provide this information, as it does not fall within basic services and may be expressly proscribed by some professional liability insurance policies.

References

Sweet, pp. 295–311.
Walker, pp. 204–241.
AIA B-2.

Performance Bond

THE AMERICAN INSTITUTE OF ARCHITECTS

AIA Document A311

Performance Bond

KNOW ALL MEN BY THESE PRESENTS: that

PHILIP PA TRENCHTN INC.

WITTS END, WISCONSIN

(Here insert full name and address or legal title of Contractor)

as Principal, hereinafter called Contractor, and

ACME GUARANTEE INC.

MILWAUKEE, WISCONSIN

(Here insert full name and address or legal title of Surety)

as Surety, hereinafter called Surety, are held and firmly bound unto

ELLEN I. WATER

P.O. BOX 314

HOLDEMAT BAY WISCONSIN

(Here insert full name and address or legal title of Owner)

as Obligee, hereinafter called Owner, in the amount of

ONE HUNDRED AND FIFTY THOUSAND DOLLARS   Dollars ($ 150,000.00   ),

for the payment whereof Contractor and Surety bind themselves, their heirs, executors, administrators, successors and assigns, jointly and severally, firmly by these presents.

WHEREAS,

Contractor has by written agreement dated FIFTEENTH MAY 1981 , entered into a contract with Owner for

(Here insert full name, address and description of project)

WATER RESIDENCE 1. LAKESIDE

HOLDEMAT BAY, WISCONSIN

in accordance with Drawings and Specifications prepared by

FAIR ¢ SQUARE A.I.A.

HOLDEMAT BAT, WISCONSIN

(Here insert full name and address or legal title of Architect)

which contract is by reference made a part hereof, and is hereinafter referred to as the Contract.

1

(page I of 4)

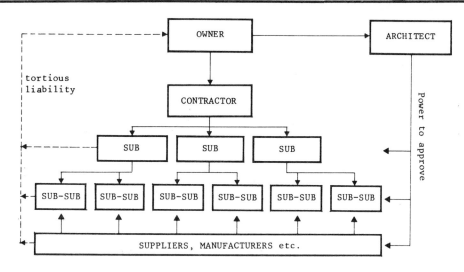

Under the single contract system, it is not unusual for prime contractors to sublet parts of the work to other contractors, either due to the size and scope of the project, or to take advantage of specialist skill or knowledge. If subcontracting is anticipated where AIA owner–contractor agreements are being used, a standard form of subcontract is advisable. The AIA produces Document A401, Contractor–Subcontractor Agreement Form, which can be used in conjunction with other AIA forms including A101, A107, A111, and A201.

The subcontract corresponds to the other AIA Documents in terms of:

- Responsibilities and liabilities
- Payment
- Relationships of parties.

The subcontractor may, in turn, delegate responsibility to other contractors who are known as sub-subcontractors. The same relationship is established as with the contractor and subcontractor, although the prime contractor still retains overall responsibility for all work undertaken on site.

Selection

The contractor may select suitable subcontractors, and cannot be forced by the owner to work with anyone to which reasonable objection can be made. However, as soon as is practicable after the owner–contractor agreement is signed, the contractor should submit to the architect a list of proposed subcontractors and suppliers. The architect and/or owner may reasonably object to any of the names on the list, but such objection should be made promptly so that the contractor may submit a substitute. If the substitution is acceptable, the contract sum can be adjusted by change order (see pages 111 and 114) to accommodate any financial inequities caused by the change. No substitution of subcontractors should be made by the contractor without architect and/or owner knowledge and approval.

Payment

Payments to the subcontractor by the prime contractor are governed by the same requirements as the owner's payments to the contractor, although the AIA agreement provides for subcontractor payment within three working days of the owner's payment, reflecting the same retainage (see page 13).

The subcontractor may request information from the architect concerning the percentage of work completed or amounts certified (under the General Conditions 9.5.3), even though no contractual relationship exists between the two parties at any time. If a certificate of payment is withheld from the contractor through no fault of the subcontractor, the latter is nonetheless entitled to payment for work completed to date, and the contractor will be bound to pay it. If the subcontractor is not paid, Article 11.12.1 of AIA Document A401 states that, after giving due notice, the subcontractor can stop work until payment is made.

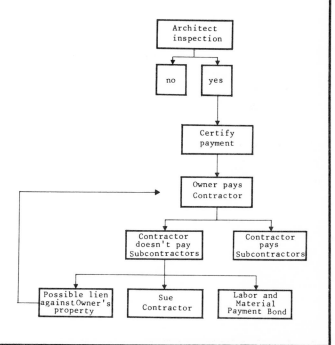

Subcontractors and Suppliers

In some cases, payment made by the owner to the contractor may not reach the subcontractor (e.g., in the event of the contractor's bankruptcy). It is possible for the subcontractor to successfully claim against the owner by filing a *mechanic's lien* against the property (see page 14), causing the owner to pay twice for the same work. To avoid such problems, labor and material payment bonds should be used to prevent undue hardship to the owner (see pages 120 and 121).

Suppliers

Material suppliers and manufacturers contract directly with the prime contractor, subcontractors, and sub-subcontractors and have no contractual relationship with either the owner or the architect. However, similar legal rights exist where nonpayment occurs, and appropriate bonds should be required from the contractor to give necessary protection.

The owner and the architect have similar rights under the AIA General Conditions A201 to reasonably object to any suppliers that the contractor intends to use.

References

Sweet, pp. 450–480.
AIA D–3.

THE AMERICAN INSTITUTE OF ARCHITECTS

AIA Document A401

SUBCONTRACT
Standard Form of Agreement Between Contractor and Subcontractor

1978 EDITION

Use with the latest edition of the appropriate AIA Documents as follows:

A101, Owner-Contractor Agreement — Stipulated Sum
A107, Abbreviated Owner-Contractor Agreement with General Conditions
A111, Owner-Contractor Agreement — Cost plus Fee
A201, General Conditions of the Contract for Construction.

THIS DOCUMENT HAS IMPORTANT LEGAL CONSEQUENCES; CONSULTATION WITH AN ATTORNEY IS ENCOURAGED WITH RESPECT TO ITS COMPLETION OR MODIFICATION

This document has been approved and endorsed by the American Subcontractors Association and the Associated Specialty Contractors, Inc.

AGREEMENT

made as of the TWENTY FIRST day of MAY in the year Nineteen Hundred and EIGHTYONE

BETWEEN the Contractor: PHILIP DA TRENCHTYN INC.
WITTS END, WISCONSIN

and the Subcontractor: WALTER WALL
HOLDEMAT BAY, WISCONSIN

The Project: WATER RESIDENCE, 1 LAKESIDE, HOLDEMAT BAY WISCONSIN

The Owner: ELLEN I. WATER

The Architect: FAIR & SQUARE A.I.A.
HOLDEMAT BAY, WISCONSIN

The Contractor and Subcontractor agree as set forth below.

AIA DOCUMENT A401 • CONTRACTOR-SUBCONTRACTOR AGREEMENT • ELEVENTH EDITION • APRIL 1978 • AIA®
©1978 • THE AMERICAN INSTITUTE OF ARCHITECTS, 1735 NEW YORK AVE., N.W., WASHINGTON, D.C. 20006

A401-1978 **1**

(page I of 7)

Action Required

1. Telephone message

For: Tom Square
Date: May 15
Time: 2:30 p.m.
From: Doug Deapleigh, contractor.
Re: Error in bid.
Taken by: Candice Courage, secretary.

Mr. Deapleigh rang to say that the bid that we accepted of his last week for the bank job contained a large error, and is underpriced by $4,000.

Can he change the bid? If not, can he withdraw it?

2. Letter

Our ref: EIW/jg
Your ref: TS/cc

May 18, 1981

Dear Sirs:

re: New House at Holdemat Bay, Wisconsin.

I have been through the contract documents you sent me, and find that I am not altogether satisfied with some of the provisions concerning the rights and responsibilities of the contractor.

I enclose a revised copy of the contract with a few amendments and additions that I think are preferable.

Yours sincerely,

Ellen I. Water

3. Memo

MEMO

To: T.S.
From: Dee Zeiner
Date: May 15, 1981
Re: New renovation job

We've been asked to deal with the renovation of that old dilapidated hotel on the coast road. The owner wants to turn it into a new restaurant and private club, and wants to get going as soon as possible. He wants to know about our fees and the best way to get it built. What do you suggest?

4. Memo

MEMO

To: Bill
From: Tom
Date: May 18, 1981
Re: Possible conflict of interest.

You know the Water job—one of the subcontractors put forward by the prime is my cousin Rocko. Should we reject him on the grounds of conflict of interest?

1. Diary insert

Date: May 20, 1981
Re: Mistaken bid.

Although contractually he can be held to the bid for 30 days, it would be better to release Deapleigh and award the contract to the next lowest responsible bidder. Allowing him to change the bid is bad practice. The difference between the bid and next lowest can be deducted from the bid bond.

2. Diary insert and letter

Diary insert

Date: May 20, 1981
Re: Ms. Water—Changes in the standard form

Write and dissuade from fooling with standardized forms—all the AIA documents we are using tend to relate to each other—if you change one, you may mess up another.

2.

The changes that you propose do not appear to merit the problems they may cause, and we assure you that the Conditions as they have been drafted are amply suited to your type of project. However, in the event that you wish to go ahead with such alterations, we feel it would be in the best interests of the project if you sought expert legal assistance in the matter.

We look forward to hearing from you,

Yours sincerely,

Fair and Square

Letter

May 20, 1981

Our ref: TS/cc
Your ref: ElW/jg

Dear Ms. Water:
re: Proposed house at Holdemat Bay, Wisconsin.

Thank you for your letter of May 17. With regard to the amendments you propose, we strongly advise that you consider the consequences of such an action. The AIA General Conditions of the Contract of Construction is a complex and comprehensive document containing many articles that are interrelated with other provisions both in the same document and in other standard forms we are using.

They have all been drafted under expert guidance to provide the best conditions for both parties concerned, and interference with the delicate relationship of articles may lead to incalculable consequences later in the project.

3. Memo attached to time sheet

MEMO

To: DEE
From: TOM
Date: May 19, 1981
Re: Old hotel renovation

The work looks tricky and we have no way of knowing what to expect either in the design or construction phases. Suggest we discuss the possibility of a negotiated contract to bring the contractor in earlier, with a cost plus award fee.

As to our fees, a Multiple of Direct Personnel Expense is preferable for all design and inspection work, but we could offer a lump sum arrangement for the construction documents phase, as we should be able to predict our work load for that stage of the work.

Have dug out time sheets in readiness, as we will need more copies. Have you set up an appointment with the owner to discuss the scheme further?

4. Memo

MEMO

To: Tom
From: Bill
Date: 5/20/81
Re: Conflict of interest with cousin

I doubt this is a problem. Just inform Ms. Water that Rocko is your cousin. If she wants to object, fine, but we will be playing it straight.

SECTION 6.
THE CONSTRUCTION PHASE

Throughout the design and construction process, a number of procedures and operations involving the architect have to be carried out which require documentation and record. As a written format is always preferable, standardized forms are useful, and many larger organizations will prepare their own personalized paperwork.

This includes letterheads, memorandum pads, and telephone message pads (see pages 39 and 41), but may also extend to more technical and detailed documents necessary in both office management and project administration.

The AIA produces a comprehensive collection of documents for use in the construction process, which are strongly recommended for their generally accepted meaning, consistency of format, and interrelated content.

The AIA documents are divided into series reflecting different aspects of administration in architectural practice:

A Series: Owner/Contractor Documents

These include all agreements designed for various construction project types and their respective conditions of agreement, bond forms, and bidding-related documents.

B Series: Owner/Architect Documents

Standard forms of agreement, duties, responsibilities, and limitations of the authority of the architect's project representative, the owner/construction manager agreement, and the architect's qualification statement are published in this series.

C Series: Architect/Consultant Documents

Forms of agreement between the architect and consultants, including the engineer, and joint venture forms.

D Series: Architect/Industry Documents

Procedures for calculating area and volume of buildings and a project checklist.

E Series: Architect/Producer Documents

Technical literature for the construction industry.

F Series: Architect's Accounting Forms—Manual System

Sheets including cash journal, payroll journal, record forms, balance sheets, expense records, project report forms, etc.

F Series: Compensation Guidelines, Forms, and Worksheets

Including worksheets indicating designated services, phase compensation, and project time and payment schedule.

F Series: Standardized Accounting for Architects Forms

G Series: Architect's Office and Project Forms

Land survey requisitions, change orders, certificates of insurance, applications for employment, and several other formats developed to assist architects in the internal running of the practice and in dealings with specific projects.

The documents are sold either individually or in units of 25 or 50. They can be obtained from regional branches of the AIA, where a discount is available for members.

The AIA also publishes the *Handbook of Professional Practice,* which is highly recommended for all practicing architects. It is published in separate volumes and contains individual chapters which are updated when necessary. Several practice aids are also published by the AIA, including:

- *The Handbook Supplement Service* (Z100)
- *The Glossary of Construction Industry Terms* (2M101)
- *AIA Building Construction Legal Citator* (2M119).

Once the building contract has been signed, the architect's role in the construction process changes, together with the architect/owner relationship. During the design development phases, the architect is seen (by some courts of law) to fulfill the role of independent contractor, whereas during the construction stage, this role becomes that of a limited agent (see page 26). The limits of this role are expressed within the owner/architect and owner/contractor agreements, and great care should be taken by the architect not to exceed or mishandle the powers necessary for the administration of the contract. If the architect's powers *are* exceeded, such acts can be ratified by the owner, but it is obviously preferable to avoid the situation if possible.

A detailed breakdown of the building contract is included on pages 79–84, although the architect's duties can be loosely grouped into three categories:

- Performance evaluation
- Certification
- Adjustment.

Performance Evaluation

A major part of the architect's work during the construction phase of a lump sum contract concerns the ensurance that the work carried out conforms to the detail and quality required by the drawings and specifications. There are no powers granted to the architect which enable him/her to tell the contractor *how* to do the work, but certain provisions within the building contract enable the architect to provide a quality control measure on behalf of the owner. These provisions can be categorized as:

- Observation
- Inspection
- Approval.

Observation (AIA Document A201, Article 2.2.5)

Contained within the provisions of the General Conditions are a number of clauses that enable the architect to undertake the required duties. The contractor agrees to allow the architect access to wherever the work is in progress, which includes workshops where components or fittings are constructed, as well as visits to the building site.

Site Visits (AIA Document A201, Article 2.2.3)

The architect should visit the site at appropriate intervals to ensure that the work is compliant with the contract documents. Frequency of visits will depend on a number of factors, including:

- Type of project
- Site conditions
- Complexity/size of project
- Stage of construction reached
- Type of owner
- Knowledge of the contractor
- Location of the site from the architect's office
- Whether an on-site architect or construction manager is being employed
- Whether additional fees are being charged for inspections
- Unforeseen events (e.g., bad weather)
- Specific events (e.g., covering-up; see page 107).

On arrival at the site, the architect should report to the contractor or the named superintendent and should communicate project matters solely with that person for the duration of the site visit. A record should be kept of all visits, noting any observations, information supplied, and actions that should be taken.

In the normal progress of the work, site visits can arise either during, at the commencement or at the completion of some of the following activities, depending upon the project:

- Establishment of datum points, bench marks, and building layout
- Dimensions and grade establishment
- Safety and security provisions
- Protection of trees or existing buildings
- Fences, hoardings, and signs
- Siting of storage areas
- Excavation and soil underfootings
- Public utility connections (telephone, gas, electricity, etc.)
- Foundations, reinforcement, pile-driving
- Concrete tests, formwork, reinforcement, and pouring
- Structural frames
- Floor openings, sleeves and hangers, floor laying
- Quality and placing of concrete
- Weather precautions
- Masonry layout and materials
- Bonding and flashing
- Frames and prefabricated elements
- Partition layout, lathing, and drywalling
- Temporary enclosures, heat, light, and sanitation during site operations
- Protection of finished work
- Fittings and cabinetwork
- Tiling, electrical work, wiring, pipework, and installation of hardware and equipment
- Roofing installation
- Painting, varnishing, and surface finish
- Equipment/plumbing tests and inspections required by public authorities.

References

Walker, pp. 54–57.
AIA 18.

Inspection

At certain stages during the construction process, the architect will appraise the work completed and issue a written judgment upon it. Appraisal is required for:

- Stage payments (AIA Document A201, Article 2.2.6)
- Substantial completion (AIA Document A201, Articles 2.2.16, 9.8.1)
- Final Inspection (AIA Document A201, Article 9.9.1)

Approvals

In addition to the above inspections or in respect of other duties required under the building contract, the architect may be called upon to make certain judgments on aspects of the work in the form of an approval or rejection (AIA Document A201, Article 2.2.13). Such instances include:

- Testing (2.2.13)
- Uncovering of work (13.1.1)
- Approvals of samples and shop drawings (2.2.14)
- Itemized values (9.2.1)
- List of subcontractors (5.2.1)
- Supporting data for payments (9.9.2).

Certification

At the stages in the construction process where inspections are carried out, it is often necessary to certify approval in writing. Such approval has the effect of releasing payment to the contractor, and should be undertaken with the greatest of care and diligence. Certification may take the form of a letter sent to both the owner and the contractor, or one of the standard forms produced by the AIA specifically for the purpose (see page 131).

Payment

In certifying payment, the architect must be satisfied that the amount of payment represents the stated value of the work (which must be reasonably accurate), less the agreed retainage and less the total of earlier certificates. In addition, the architect should be sure there is nothing to prevent the certificate being granted (e.g., defective work not remedied A201, Article 9.5.1.) and require such evidence substantiating the contractor's right to payment as considered necessary (Article 9.3.1).

Adjustment

This category of architectural service is the least defined in the contract documents, although provisions for its implementation can be found throughout the General Conditions. Basically, the powers connected with this category ensure that, in the event of confusion or disagreement between the owner and contractor, or in the event of unforeseen changes or conditions occurring, the architect can act to maintain the continued progress and quality of the work. The architect's duties in this respect fall into two main categories:

- Interpretation
- Modification.

Interpretation

Article 2.2.8 of the General Conditions gives the architect authority to render interpretations of the intent of the contract documents in the event of the parties failing to agree. This helps to solve any ambiguities, and to keep minor disagreements or unclear requirements from delaying the project. Either the owner or the contractor can require the architect to interpret an aspect of the contract documents, and the architect should make the decision in writing within a reasonable time (A201, Article 2.2.9).

In the role of interpreter, the architect is expected to act as arbitrator and "to secure faithful performance by both the Owner and Contractor". To help the architect assume an unbiased position in this, a quasi-arbitral immunity is granted for decisions made under this provision, removing any liability for the results of the interpretation, rendered in good faith by the architect. Consequently, the architect should undertake interpretive duties with a totally unbiased attitude, and not allow employment ties to the owner to affect the outcome of decisions.

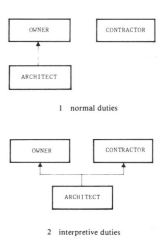

1 normal duties

2 interpretive duties

Modification

Where circumstances or new requirements mean that the contract documents need to be amended, the architect is empowered to make certain minor changes (Article 12.4.1, see page 111) or issue (but not approve) change orders permitting new work to be undertaken (Article 2.2.15). Other actions in this respect may be carried out if prompted by the acts or omissions of either party (e.g., acceptance of nonconforming work by the owner, Article 13.3.1).

Progress Appraisal

A proportion of the architect's duties during the construction phase concerns the checking of the work to ensure that it will be completed by the agreed date of completion.

Several mechanisms may be used by the architect to monitor building progress throughout the project. These include:

- Site visits and reports
- The contractor's work schedule
- Schedule of values
- Meetings.

Site Visits (see page 106)

Visits to the building site should be made "at intervals appropriate to the stage of construction" to familiarize the architect with the progress and quality of the work. Observations made during these visits should be recorded and copies sent to the parties involved, who may include:

- The owner
- Consultants
- Field architect, if appropriate
- Construction manager, if appropriate.

Although any type of record will be sufficient for noting the outcome of site visits, standardized formats are recommended for the sake of consistency and conformity of files. The AIA publishes Document G711, Architect's Field Report (see page 110), which provides categories to note and comment upon the following:

- The stage of completion
- Temperature, weather
- Date
- Work in progress

- Persons present
- Conformance with schedule
- Any observations
- Items to verify
- Information or action required.

The Contractor's Work Schedule (see page 91)

This schedule represents the contractor's intended plan of work established at the outset of the construction phase. Comparison between the projected progress and actual advancement of the work provides a means of assessing the overall conformity of the project to the original timetable.

Schedule of Values

Similarly, the schedule of values which allocates value to various amounts or portions of the work can be used to a lesser degree to establish how well the original estimates of cost allocation match up to actual certification.

Meetings

Meetings between various parties concerned with the construction process may be held periodically. Types of meetings include:

- Practice meetings
- Contractor meetings
- Site meetings.

Practice Meetings

Meetings between partners and/or employees may be held at intervals to discuss either practice policy or a specific project that is in progress or scheduled to begin.

Contractor Meetings

The contractor and representatives may wish to meet with the subcontractors at intervals to discuss coordination of work on site. The architect or construction manager may be invited to attend where relevant.

Site Meetings

Meetings between parties representing different elements of the construction process may be necessary at intervals throughout the project.

They may be held:

- At regular intervals
- At specific times during the construction process
- When problems occur
- When it seems necessary to provide an impetus.

Those attending, in addition to the architect and the contractor (and/or the project representative) may include:

- The owner
- The construction manager (if one is employed)
- Consultants
- Subcontractors
- Others (e.g., the building inspector).

However, the nature of the project will largely determine the make-up and nature of the meeting.

Procedure

Although there is no standardized format involved in setting up and running meetings, certain basic guidelines are suggested for adoption.

Whoever takes responsibility for chairing the

meeting—and this role may be taken by the construction manager, the contractor, or the architect—is likely to prepare and distribute the minutes of the previous meeting and notify parties of the next one.

Parties should be notified well in advance of the time and date of the proposed meeting, and sent a copy of the previous meeting's minutes for consideration and file. Parties unable to attend should notify the Chair of their situation as soon as possible so that, in the event of their presence being necessary at the meeting, a new date may be scheduled which is amenable to all concerned.

The Agenda

The agenda of a typical site meeting might be set out in the following way.

The Chair should:

- Call the meeting to order
- Take the names of those present
- Give the names of those sending apologies for their absence.

The rest of the meeting might then include:

- Agreeing the minutes of the last meeting, or dealing with any problems arising from them
- The architect's report
- The construction manager's report
- The contractor's report
- Any consultants' reports
- Discussion of project progress
- Procedures and communications necessary (any actions required, by whom, etc.)
- Any other business
- Time and place of next meeting.

Architect's Field Report

ARCHITECT'S FIELD REPORT

AIA DOCUMENT G711

OWNER ☒
ARCHITECT ☒
CONSULTANT ☐
FIELD ☐

PROJECT: WATER RESIDENCE J. LAKESIDE HOLDEMAT BAY   FIELD REPORT NO: 4

CONTRACT: SINGLE FAMILY DWELLING   ARCHITECT'S PROJECT NO: 203

DATE 10 AUG 81   TIME 11:00 AM   WEATHER CLEAR   TEMP. RANGE 50-60

EST. % OF COMPLETION 70%   CONFORMANCE WITH SCHEDULE (+, −) +

PRESENT AT SITE PHILIP DA TRENCHYN, P. TAYLOR, I. SAWYER

WORK IN PROGRESS   ELECTRICAL
PIPE INSULATION
CARPENTRY

OBSERVATIONS   BATHROOM FITTINGS INSTALLED. 8 DAMAGED TILES TO BE REPLACED TO
WEST WALL BELOW WINDOW.
KITCHEN CABINETS IN PLACE ALTHOUGH NOT COMPLETED. SOME DAMAGE TO TRIM
ON TWO UNITS.
PIPE INSULATION IN ROOF VOID CHECKED AND VERIFIED FOLLOWING INSPECTION
SPECIFIED PAINT TO WOOD TRIM TO EXTERIOR WINDOW FRAMES CONSIDERED UNSUITABLE
COLORING AND NEEDS TO BE CHANGED TO A LIGHTER HUE BEFORE CONTRACTOR HAS
IT DELIVERED

ITEMS TO VERIFY   NONE

INFORMATION OR ACTION REQUIRED   CHECK WITH OWNER WHETHER ROCKS ON SITE SHOULD BE
STACKED FOR OWNERS USE OR REMOVED. CHANGE ORDER REQUIRED FOR ADDITIONAL
ROOF ACCESS. SUPPLEMENTAL CHANGE ORDER FOR PAINT REPLACEMENT

ATTACHMENTS

REPORT BY:

AIA DOCUMENT G711 • ARCHITECT'S FIELD REPORT • OCTOBER 1972 EDITION • AIA® • © 1972
THE AMERICAN INSTITUTE OF ARCHITECTS, 1735 NEW YORK AVE., NW, WASHINGTON, D.C. 20006

page 1 of 1 page

110

During the construction phase, it may become necessary to amend the original contract documents by addition, alteration, or deletion as a result of:

- Unforeseen or unexpected events
- New requirements
- New circumstances invalidating parts of the contract documents.

The AIA General Conditions provide for changes to be made, but care should be taken to identify the nature of the change sought, and to deal with it in the appropriate manner. Changes may fall into the following categories:

- A modification
- A cardinal change
- A constructive change
- A change
- A minor change
- Other.

Modifications

At any time during the contract period, the owner and contractor may mutually agree to change the intent or substance of the contract between them. As the contract is a voluntary agreement between the parties, any joint acquiescence as to its content is acceptable (see page 73), but great care should be taken in the modification of documents and the revised provisions for payment, work definition, etc.

Cardinal Changes

If the owner demands a change in the contract documents which goes beyond the intent of the original contract, this may be construed as a major change or, to use the federal procurement expression, a "cardinal change." Such a change may give the contractor sufficient justification to stop work and to claim damages for the owner's breach of contract. In privately funded projects, the contractor may wish to renegotiate payment for a new contract, whereas in publicly sponsored work, it may be necessary to readvertise the project.

Constructive Changes

"Constructive changes" are referred to in federal procurement projects, and occur when the contractor is asked to undertake work:

a. which is different from that required by the contract
b. which speeds up the project
c. which requires added expenditure as a result of incorrect specifications.

If forced to make a constructive change, the contractor may require the contract sum to be adjusted accordingly.

Changes

If the AIA contract is used, as long as changes required by the owner are within the general scope of the contract, the contractor will be required to undertake the work, with or without the latter's consent. The contract time and the contract sum may be adjusted to compensate for the extra work. Payment for changed requirements could be:

- By mutual agreement on a lump sum
- By unit prices (either agreed upon, or previously stated in the contract documents)
- By an agreed cost of the work plus a fixed or percentage fee
- By determination of the architect.

Ordering a Change

Prior to requiring a change, it is often advisable to establish the final cost of the work involved. AIA Document G709, Proposal Request, may be sent to the contractor to ask for an account of the increased cost and/or time that will be necessary. If the owner decides to continue with the changed requirements, the architect will prepare and sign a change order (see page 114) and send it to the owner for signing before passing it on to the contractor.

Under the AIA General Conditions, the Change Order (AIA Document G701) is the only acceptable means by which the contract time or the contract sum may be altered. When signing the change order, the contractor indicates agreement with the proposed changes and becomes entitled to any justifiable extra payment (see page 116).

In the event that time is of the essence in the contract, and to prevent delay due to the administrative procedures involved, the process may be expedited by the use of AIA Document G713, Construction Change Authorization. This is not a change order, but an authorization to the contractor to proceed with the work prior to the issuance of the change order which will be "promptly processed".

Minor Changes

When alterations to the contract documents are considered necessary, but are sufficiently small so as not to change the contract time or the contract sum, the architect is empowered to order such alterations which are referred to as "minor changes" (AIA Document A201, Article 12.4.1). Both the owner and the contractor will be bound by such written orders which are usually issued on AIA Document G710, Architect's Supplemental Instructions (see page 113).

Contract Changes

Other Forms of Change

Due to the unpredictable and complex nature of many building projects, certain changes are sometimes necessary to provide for specific contingencies which include:

- Emergencies (AIA Document A201, Article 10.3.1)
- Concealed conditions (AIA Document A201, Article 12.2.1)
- Escalation.

Emergencies

If the safety of persons or property is threatened in any way, the contractor may act at his/her discretion to prevent loss or injury. The architect may then determine the effects of the emergency on the project, and reflect them in a subsequent change order regarding contract time and contract sum.

Concealed Conditions

Because of the unpredictable nature of subsurface conditions, some contracts provide remedies to equitably adjust the contract sum and time if conditions prove to be materially different from those anticipated. The AIA General Conditions contain such provision, requiring that claims in respect of concealed conditions by either party be made within 20 days of their detection.

Escalation and Fluctuation

Inflation and price escalation may make the estimation of a stipulated sum price difficult, possibly causing the contractor to overbid to protect against financial loss by erosion of profit. It is possible to add to any contract a fluctuations clause which provides an agreed method of calculation in the event of sudden price variations.

References

Sweet, pp. 319–320, 346–368.
AIA 18.

**ARCHITECT'S
SUPPLEMENTAL INSTRUCTIONS**

AIA DOCUMENT G710 (Instructions on reverse side)

Owner ☒
Architect ☒
Consultant ☐
Contractor ☒
Field ☐
Other ☐

**ARCHITECT'S SUPPLEMENTAL
INSTRUCTION NO:** 3

DATE OF ISSUANCE: 13TH AUG 1981

ARCHITECT: FAIR & SQUARE A.I.A.

PROJECT: WATER RESIDENCE 1. LAKESIDE
(name, address) HOLDEMAT BAY, WISCONSIN

OWNER: ELLEN I. WATER

TO: PHILIP DA TRENCHYN
(Contractor)

ARCHITECT'S PROJECT NO: 203

CONTRACT FOR: SINGLE FAMILY DWELLING

The Work shall be carried out in accordance with the following supplemental instructions issued in accordance with the Contract Documents without change in Contract Sum or Contract Time. Prior to proceeding in accordance with these instructions, indicate your acceptance of these instructions for minor change to the Work as consistent with the Contract Documents and return a copy to the Architect.

Description: SUBSTITUTE EXTERNAL FINISH B41 OVERALL LO-SHEEN FINISH

16 LINE FOR SPECIFIED XXZ OVERALL LATEX HI-GLOSS ENAMEL

43 LINE TO ALL EXTERNAL WINDOW FRAMES WHERE SPECIFIED

Attachments: *(Here insert listing of documents that support description.)*

NONE

ACCEPTED: 8.15.81

BY *P da Trenchyn* 8.16.81
   Contractor   Date

G710 — 1979

ISSUED: 8.13.81

BY *B. Fair*
   Architect

AIA DOCUMENT G710 • ARCHITECT'S SUPPLEMENTAL INSTRUCTIONS • MARCH 1979 EDITION • AIA®
©1979 • THE AMERICAN INSTITUTE OF ARCHITECTS, 1735 NEW YORK AVE., N.W., WASHINGTON, D.C. 20006

Change Order

CHANGE ORDER

AIA DOCUMENT G701

Distribution to:
- ☒ OWNER
- ☒ ARCHITECT
- ☒ CONTRACTOR
- ☐ FIELD
- ☐ OTHER

PROJECT: WATER RESIDENCE
(name, address) 1, LAKESIDE
HOLDEMAT BAY

TO (Contractor):
PHILIP DA TRENCHYN INC.
WITTS END, WISCONSIN

CHANGE ORDER NUMBER: 2

INITIATION DATE: AUGUST 15TH 1981

ARCHITECT'S PROJECT NO: 203

CONTRACT FOR: SINGLE FAMILY DWELLING

CONTRACT DATE: 15 MAY 1981

You are directed to make the following changes in this Contract:

CONSTRUCT AND INSTALL CEILING HATCH AND COLLAPSIBLE STAIR TO CEILING OF
HALLWAY AS SHOWN IN DRAWING 107 ENCLOSED

COPY

Not valid until signed by both the Owner and Architect.
Signature of the Contractor indicates his agreement herewith, including any adjustment in the Contract Sum or Contract Time.

The original (Contract Sum) (Guaranteed Maximum Cost) was .... $ 150,000.00
Net change by previously authorized Change Orders .... $ 780.00
The (Contract Sum) (Guaranteed Maximum Cost) prior to this Change Order was .... $ 150,780.00
The (Contract Sum) (Guaranteed Maximum Cost) will be (increased) (decreased) (unchanged)
  by this Change Order .... $ 1,240.00
The new (Contract Sum) (Guaranteed Maximum Cost) including this Change Order will be .... $ 152,020.00
The Contract Time will be (increased) (decreased) (unchanged) by
The Date of Substantial Completion as of the date of this Change Order therefore is 27TH DECEMBER 1981 (4) Days.
Authorized:

ARCHITECT FAIR & SQUARE A.I.A.
Address HOLDEMAT BAY, WISCONSIN
BY B Fair.
DATE 8.15.81

CONTRACTOR PHILIP DA TRENCHYN
Address WITTS END, WISCONSIN
BY P da Trenchyn
DATE 8.18.81

OWNER ELLEN I. WATER
Address P.O. BOX 314, HOLDEMAT BAY WI
BY E. J. Water
DATE 8.22.81

114

Many stipulated sum building contracts are drafted on the basis that time is an important factor. The AIA General Conditions, for example, are drafted to include the provision that "time is of the essence" (Article 8.2.1). Such provisions make it important for the contractor to complete the work in conformance with the contract documents, on or before the date of substantial completion stipulated in the contract.

If the contractor fails to finish within the specified time, the contract is breached and several mechanisms may come into effect as a result, such as:

- Liquidated damages
- Termination
- Refusal of further payment
- Variations
- Extensions of time.

Liquidated Damages

These basically represent a pre-agreed formula that can be used as a basis of penalty against the contractor for late work. They are usually determined as a fixed sum per day, payable for every working day beyond the date of substantial completion. The extent of the financial amount involved for each day's delay will depend upon how critical prompt completion is considered by the owner.

The requirement is generally expressed in the bidding documents, and may affect the contractor's bid. If very high penalty clauses are used, they are often supplemented with a corresponding bonus clause. The bonus is often expressed as the same amount as the penalty and is payable to the contractor for every day saved prior to the substantial completion date.

Termination

In certain extreme circumstances of delay, there may be justification to terminate the contract between the owner and the contractor (see page 137).

Refusal of Further Payment

In some cases the contractor may be denied further payment. However, this should be handled carefully, as it may provide grounds for termination on the part of the contractor.

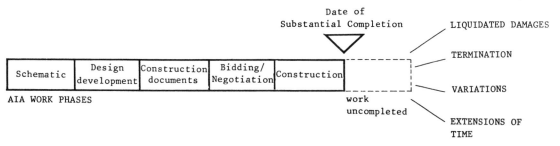

Note: Delay on the part of the contractor need not necessarily result in a penalty if sufficient cause can be shown to substantiate a legitimate alteration to the contract documents.

Variations

If changes are made to the contract requirements by the owner, the architect can issue a change order which may provide for extra payment to the contractor, as well as extra time for completion (see page 111).

Extensions of Time

In some cases, unforeseen or unavoidable occurrences will delay the progress of the work through no fault of the contractor. The AIA General Conditions provide for extensions to the contract time to be granted by the architect for delays caused by:

- Act or neglect of the owner or architect (or employee of either)
- Act or neglect of a separate contractor (but *not* subcontractor; see page 85)
- Changes ordered to the work
- Labor disputes
- Fire
- Unusual delay in transportation
- Adverse weather conditions (not reasonably anticipatable)
- Unavoidable casualties
- Any cause beyond the contractor's control (or acts of God, including: earthquake, landslide, hurricane, tornado, lightning, flood, etc.)
- Delays authorized by the owner pending arbitration
- Any other cause that the architect determines to be justifiable.

Claims for Extensions

Should the contractor feel that an extension is warranted, an application must be made in writing to the architect within 20 days of discovery of the event likely to cause delay. An indication of the probable effect of the delay upon the construction work should be included, and the contractor should be encouraged to make every reasonable effort to minimize the impact of the event on the general progress of the project.

The granting of an extension is not automatic; for example, bad weather alone may be insufficient to warrant extra time. It must be shown (e.g., by reference to meteorological records) that the weather in question was far worse than the norm for the time of year, and actually delayed operations on site.

Similarly, a claim for delay due to labor disputes may be disallowed if the dispute was in progress at the time of contract formation. The onus, therefore, is on the contractor to show both the justifiable reason for an extension, and its impact upon the progress of the work.

If there is sufficient cause to justify the granting of an extension, there may also be grounds for additional compensation. Such claims often arise from owner delay and decision, and may include:

• Site not ready in time for contractor occupation
• Delays in progress payments
• Delays in issuing change orders
• Delays in approving submittals
• Errors in drawings and/or specifications
• Administrative delays (poor coordination of separate contractors, inspection delays, etc.).

Claims can be made by the contractor to compensate for:

• Labor costs (including subcontractor costs, wages, overtime, insurance, etc.)
• Equipment costs
• Material costs (additional and escalation costs)
• Overhead (field and office)
• Insurance and bond costs
• Other losses (seasonal problems, congestion on site, etc.).

Reference

Sweet, pp. 390–416, 481–506, 600–638.

Other factors relating to the time element of the construction process include:

- Acceleration
- Stopping the work
- Impossibility.

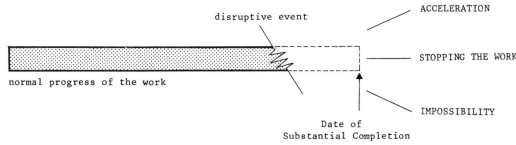

Acceleration

This can be defined in two ways:

1. Actual acceleration
2. Constructive acceleration.

1. Actual Acceleration

This may take place if the contractor is requested to complete the work *before* the date established in the contract documents. Actual acceleration is at the contractor's discretion and may provide the basis for increased costs.

2. Constructive Acceleration

This is less clear in its definition, and may occur where the contractor has experienced delay, but has not been granted an extension of time. Consequently, the contractor must make up the lost time in order to finish by the agreed date, and effectively accelerate the pace of the work.

Although acceleration was essentially a federal procurement matter (see page 111), it is now becoming more applicable to private construction contracts.

Stopping the Work

Under certain circumstances, the work may be stopped, as opposed to delayed. This may be as a result of:

- Owner's instructions
- Circumstances forcing a construction suspension.

Owner's Right to Stop the Work

AIA General Conditions (Article 3.3.1) provide the owner with the right to stop the work:

a. If the contractor fails to correct defective work
b. If the contractor refuses to carry out the work in conformance with the contract documents.

The architect does *not* have the power to stop the work unless expressly authorized to do so by the owner in writing. Previous editions of the AIA contract included the architect's right to stop the work, but this has been discontinued in the 13th edition.

Constructive Suspension

This may take place:

- If the contractor is not paid
- If notices and/or information are delayed excessively
- If change orders are delayed
- If certificates are unreasonably withheld or delayed
- If the construction documents are defective.

Constructive suspension allows the contractor to stop work (AIA Document A201, Article 9.7.1), and to claim extra compensation for the costs involved in shut-down, delay, and recommencement. Some suspensions may eventually lead to the termination of the construction contract by either party (see page 137).

Impossibility

A further reason which may be claimed as the cause of delay, and possibly lead to the termination of the contract, is impossibility of completion. If sufficient cause exists to prove that the work cannot be finished, the contractor may be excused from further performance, and might be able to recover damages from the owner.

Impossibility of completion is generally classified as either:

1. Actual
2. Practical

1. Actual Impossibility

This arises when events occur which actually prevent performance from taking place (e.g., acts of God, or determination by a governmental department).

2. Practical Impossibility

In this case, completion of the work is technically possible, but only at excessive cost due to subsequent events making the original contract sum inadequate. If it is unreasonable for the contractor to assume the higher costs, or where it would cause excessive difficulty, loss, or possible damage, the parties may be released from their contractual obligations.

APPLICATION AND CERTIFICATE FOR PAYMENT

TO (Owner): ELLEN I. WATER
P.O. BOX 314
HOLDEMAT BAY, WISCONSIN

PROJECT: WATER RESIDENCE 1. LAKESIDE
HOLDEMAT BAY, WISCONSIN

APPLICATION NO: I

PERIOD FROM: AUG. 2, 1981
TO: SEPT. 2, 1981

Distribution to:
☒ OWNER
☒ ARCHITECT
☒ CONTRACTOR
☐
☐

ATTENTION: ELLEN I. WATER

CONTRACT FOR: SINGLE FAMILY DWELLING

ARCHITECT'S
PROJECT NO: 203

CONTRACT DATE: 15 MAY 1981

CONTRACTOR'S APPLICATION FOR PAYMENT

CHANGE ORDER SUMMARY		
	ADDITIONS	DEDUCTIONS
Change Orders approved in previous months by Owner		
TOTAL		
Approved this Month		

Number	Date Approved		
1	7/17/81	$ 780.00	
2	8/15/81	$ 1,240.00	
TOTALS		$ 2,020.00	
Net change by Change Orders	$ 2,020.00		

The undersigned Contractor certifies that to the best of his knowledge, information and belief the Work covered by this Application for Payment has been completed in accordance with the Contract Documents, that all amounts have been paid by him for Work for which previous Certificates for Payment were issued and payments received from the Owner, and that current payment shown herein is now due.

CONTRACTOR:

By: _P da Tranchyn_ Date: 9.1.81

Application is made for Payment, as shown below, in connection with the Contract.
Continuation Sheet, AIA Document G703, is attached.

The present status of the account for this Contract is as follows:

ORIGINAL CONTRACT SUM .... $ 150,000.00

Net Change by Change Orders .... $ 2,020.00

CONTRACT SUM TO DATE .... $ 152,020.00

TOTAL COMPLETED & STORED TO DATE .... $ 26,850.00
(Column G on G703)

RETAINAGE __10__ % .... $ 2,685.00
or total in Column I on G703

TOTAL EARNED LESS RETAINAGE .... $ 24,165.00

LESS PREVIOUS CERTIFICATES FOR PAYMENT .... $ —0—

CURRENT PAYMENT DUE .... $ 24,165.00

State of: WISCONSIN   County of: MEDFLY
Subscribed and sworn to before me this 2ND day of SEPT., 1981
Notary Public:
My Commission expires: JUNE 1983

ARCHITECT'S CERTIFICATE FOR PAYMENT

In accordance with the Contract Documents, based on on-site observations and the data comprising the above application, the Architect certifies to the Owner that the Work has progressed to the point indicated; that to the best of his knowledge, information and belief, the quality of the Work is in accordance with the Contract Documents; and that the Contractor is entitled to payment of the AMOUNT CERTIFIED.

AMOUNT CERTIFIED .... $ 24,165.00
(Attach explanation if amount certified differs from the amount applied for.)
ARCHITECT: FAIR & SQUARE A.I.A.

By: _B. Fair._ Date: 9.2.81.

This Certificate is not negotiable. The AMOUNT CERTIFIED is payable only to the Contractor named herein. Issuance, payment and acceptance of payment are without prejudice to any rights of the Owner or Contractor under this Contract.

Action Required

1. Letter

August 10, 1981

Dear Mr. Square,

I am writing to you in connection with the warehouse job at Witt's End, where we are working as subcontractors.

We have completed most of our part of the work now, but have not received any payment for over five weeks. Attempts to contact the contractor's accounting office have been fruitless, and his supervisor is evasive on the subject. Can you help us out please?

Looking forward to hearing from you,

Yours sincerely,
Walter Wall

P.S. I believe Acme Supplies, the carpeting suppliers are having similar problems.

2. Telephone message

To: Tom
From: Ellen I. Water
Date: 8/10/81
Time: 4:30 p.m.
Re: Alterations to work in progress
Taken by: Candice Courage

Ellen I. Water called to say she visited the site yesterday, and decided that she wants:

  a. an extra bay window added to the master bedroom

  b. an extra wing added to the boundary (east) side for guests.

Will you sort this out with her please?

3. Letter

Our ref: PdT/am   August 11, 1981

Dear Sirs,
  re: House for Ellen I. Water, Holdemat Bay

In accordance with the contract documents, we are requesting a three week extension to the contract time to allow for the last period of bad weather and problems caused by subcontractor delay and wildcat strikes. Would you alter the Date of Substantial Completion accordingly?

Yours faithfully,

Phil da Trenchyn
Prime contractor

4. Journal article and memo

Journal article

"DOUBLE PAYMENT" THREAT TO OWNERS

Recent case law indicates that, despite the lack of a contractual relationship between the owner and the subcontractor or supplier, there is a possibility of a successful claim against the former by either of the latter parties in the event of their nonpayment by the prime contractor. If work undertaken by the subcontractor (or materials supplied by the supplier) forms part of work approved in a certificate of payment and such payment is made to the contractor who, for whatever reason, neglects to pay the others, the subcontractor (or supplier) may file for a mechanic's lien against the property of the owner. This will have the effect of making the owner pay twice for the same work in order to have the lien released from the property.

MEMO

To: D. Taylor
From: Tom
Date: Aug. 11, 1981
Re: Recent site visit to Water house.

The notes from the site visit we carried out yesterday are on my desk. Would you please fill out the relevant forms ready for signing for when I get back from Chicago?

5. Memo

MEMO

To: Bill Fair
From: Tom Square
Date: Aug. 12, 1981
Re: Double payment by owner threat

Saw this in the journal—could we be involved if we didn't tell the client of the risk?

Memo

1. Diary insert

To: Tom
From: Bill
Date: August 14, 1981
Re: Nonpayment of subcontractor, Walter Wall

Checked that our certificate was issued on time—the contractor should have paid within 3 days of payment.

There's not much we can do as it is a matter between the contractor and the subcontractor. If requested, we can tell Wall the percentage of completion of amount certified for his work (A401, 12.4.2). We could also lean on the contractor a little and remind him of his responsibilities—remember we can withhold certificates for nonpayment of subcontractors and suppliers A201 (9.6.13) and the contract can even be terminated (14.2.1). Remember to ask for evidence before the certificate((9.6.1), and keep an eye on the general progress of the work.

Could you acknowledge Wall's letter, but make no offers. We can do even less for Acme Supplies.

2. Letter

Our ref: TS/cc
Your ref: EIW/jg   August 11, 1981

Dear Ms. Water:
    Re: House in Holdemat Bay, Wisconsin

We received your telephone message yesterday regarding a new bay window to the master bedroom and an extra wing to the east boundary side of the house.

Before proceeding with these changes, it will be necessary to establish:
  a. if any further governmental approvals or notifications are required
  b. whether or not the additions are sufficiently within the scope of the contract to be handled under a change order.

When we have established these factors, we will be able to advise you as to the additional costs involved in the work, including our additional services, and the extra time necessary beyond the present date of substantial completion.

Information regarding the implications of the extra work will be sent to you as soon as practicable. If this is acceptable to you, please send us written notification to proceed with the work.

                    Yours sincerely,

                    Fair and Square

3. Letter and diary insert

Letter

Our ref: TS/cc
Your ref: PdT/am   August 17, 1981

Dear Mr. da Trenchyn:
    re: Water House, Holdemat Bay

We acknowledge receipt of your letter of August 11, the contents of which we note.

Prior to any action on our part regarding the extension of time required, we will need further information concerning the basis of your claim, with details of the adverse weather conditions and labor disputes. Delays caused by actions of your subcontractors will not be sufficient grounds for an extension.

We will also require an estimate of the probable effect of the alleged delays on the progress of the work.

We are confident that you will use your best efforts to minimize the effects of any delays upon the work, and assure you of our prompt attention as soon as we receive the relevant information.

                    Yours sincerely,

                    Fair and Square

4. Memo

To: Tom
From: D. Taylor
Date: Aug. 17, 1981
Re: Site visit paperwork

I have filled out the forms relating to the site visit to the Water House. Please check them and sign—they are on your desk.

MEMO

To: all personnel
From: B.F.
Date: Aug. 14, 1981
Re: double payment by owner threat

The threat of "double payment" expressed in the article now pinned to the notice board can be avoided by advising the use of labor and material payment bonds. In this way, should a subcontractor or supplier be denied payment already paid to the prime, relief can be sought from the surety. Be sure to advise all clients of the advisability of this procedure, and the dangers of not using bonds. We cannot demand their use and *must not* give detailed advice on either bonds or insurance (our E and O policy doesn't allow it). However, suggesting the client seeks professional assistance in their usage clears us of possible claims for not doing our duty.

Note: Where relevant, always ask for proof of payments from the contractor before issuing the next certificate of payment (9.3.1.).

5. Memo

Date: August 14, 1981
Re: claim for extension, Phil da Trenchyn

Need more information before granting (meteorological report, etc.). Was the dispute causing the strike expected or going on at the time of contract formation—must check these things out.

Delays by subcontractors are the contractor's responsibility—not allowed.

Diary insert

SECTION 7.
COMPLETION

Contents   **Page:**

When the contractor is nearing completion of the work, a number of procedures are recommended to ensure smooth completion of contractual performance. These procedures are the same for full completion and partial completion, where a designated portion of the work may be ready for occupation.

Substantial Completion

As soon as the contractor decides that the project has reached a state of substantial completion (i.e., when the owner can occupy or utilize the work for its intended purpose), a *punch list* is prepared. The punch list contains details of all outstanding items that the contractor intends to complete or correct, and it is sent to the architect who may then make arrangements for inspection.

The architect can amend the punch list and add extra items, if necessary. In the event that the architect feels that the work is *not* substantially complete, the contractor will be informed, and the architect need not return to reinspect until sufficient evidence is available to suggest that the work has reached the required standard. When the architect's inspection indicates that substantial completion has been reached, the Certificate of Substantial Completion will be issued (AIA Document G704, see page 131). This is prepared by the architect and sent to the owner and the contractor. The Certificate of Substantial Completion is an important document which has an effect upon:

- The contractor's warranty period
- The architect's liability period (in some instances; see page 11)
- The responsibilities of the owner and the contractor in respect of:

Site security
Insurance
Heat and utilities
Damage to the work
Maintenance.

The architect should take care when issuing the certificate to ensure that the work has in fact been substantially completed. The certificate will establish the date of substantial completion and indicate the time allowed to complete the outstanding work. Following receipt of the certificate, the contractor can apply for payment in the normal way (see page 107). This payment should take account of the retainage agreed upon in the contract documents.

Final Completion

When the items on the punch list have been completed, the contractor should notify the architect in writing. The architect must then promptly inspect the work and, if it appears to be in conformance with the contract drawings and specifications, will issue a final certificate for payment, which is usually in the form of a letter. Upon issuance of the final certificate, the contractor becomes entitled to payment for all outstanding sums. However, certain state lien laws may make it desirable to withhold a percentage of the retainage for a period of time. If this is considered necessary, it should be stated in the bidding documents and in the owner-contractor agreement.

Before final payment is made, the owner and architect should carefully check that:

- All required certificates of inspection, bonds, record drawings, and warranties have been delivered to the owner

- Keying schedule delivered (if not already undertaken)
- All work has been completed in conformance with the contract documents
- Any instructions regarding operation of equipment have been supplied
- All accounts are adjusted (contract sum, deductions, change orders, deductions for uncorrected work, AIA A201, 13.3.1).

Before the final payment is made, it is also usual to take certain safety measures. These include:

- Ensuring that the owner is protected from all possible lien claims (AIA Document G706; see page 128)
- Requiring an affidavit that all wages, and bills for materials and equipment (or other debts connected with the work which might conceivably revert to the owner) are paid in full.

If any payments by the contractor are still outstanding, the owner may require indemnification against third party claims. As an added precaution, the consent of the the contractor's surety should be obtained prior to final payment (AIA Document G707; see page 130). Should the retainage be released or reduced at any time, AIA Document G707a, Consent of Surety to Reduction in or Partial Release of Retainage, may be used.

References

Sweet, pp. 391–393, 430–434.
AIA D-3.

Completion 2

Additional safety measures may be taken by the owner depending upon:

- The contractor's reputation
- Customs and practices of the area
- State lien laws
- Owner requirements.

If there are any exceptions to the normal procedures recommended by the AIA relating to final safeguards, a payment or bond by the contractor can be used to discharge further responsibility.

Final Payment

Final payment by the owner of the balance of the contract sum, plus any remaining retainage, constitutes a waiver of all claims against the contractor except for:

- Unsettled liens
- Faulty or defective work appearing after substantial completion
- Failure of compliance with the contract documents
- The terms of any special warranties that may have been provided.

Similarly, aceptance of the final payment by the contractor waives all rights to any further claims against the owner, with the exception of any claims made in writing at the time of the final application for payment.

Post-Completion Services

At the completion of the construction work, the architect submits a final account to the owner for outstanding payment. Any work undertaken beyond this time forms the basis for additional compensation, and may include:

- Furnishing a set of amended "as-built" drawings for the owner's records, which may be useful if further work or adaptation are anticipated.
- Site visits and advice to the owner concerning work to be undertaken by the contractor during the 12-month warranty period.
- Inspection of the project prior to the expiration of the warranty period, and compilation of a report listing necessary repairs or corrections. An inspection may be arranged by the architect to check the work when complete.
- Maintenance advice or reports may be undertaken, possibly on the basis of an annual retainer.
- Post-occupancy evaluations may be carried out once the building is in use to judge its success and general performance.

Continuing Liability

At the completion of the project, the contractual relationship between the architect and the owner comes to an end. However, a period of continuing liability follows under the law of tort, during which time the architect may still be liable for negligent acts or omissions. The length of this continuing liability period is established by individual state law, and may vary considerably (see page 11).

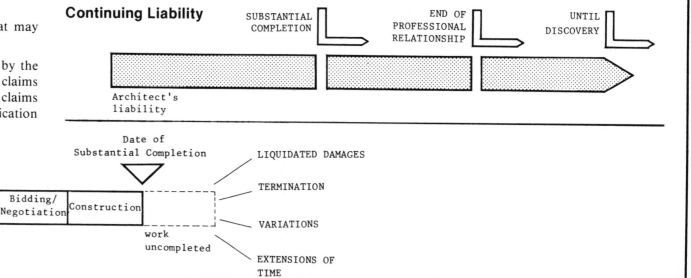

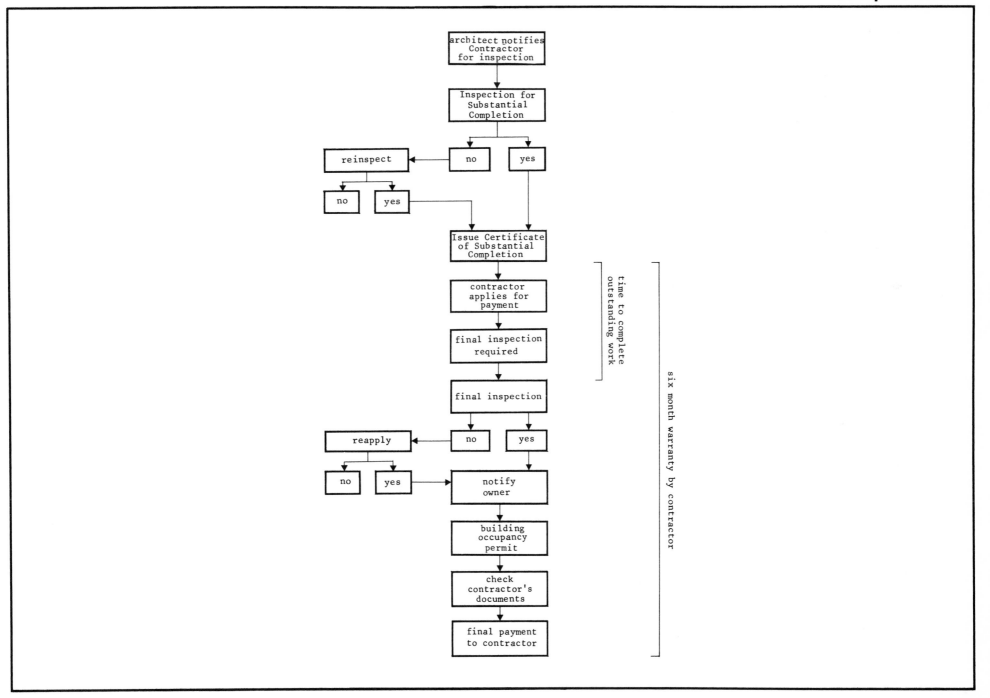

Contractor's Affidavit of Payment of Debts and Claims

CONTRACTOR'S AFFIDAVIT OF PAYMENT OF DEBTS AND CLAIMS

AIA Document G706

OWNER	☒
ARCHITECT	☒
CONTRACTOR	☒
SURETY	☐
OTHER	

ARCHITECT'S PROJECT NO: 203

CONTRACT FOR: SINGLE FAMILY DWELLING

CONTRACT DATE: 15 MAY 1981

TO (Owner)

⌐ ELLEN I. WATER
 P.O. BOX 314
 HOLDEMAT BAY, WISCONSIN ⌐

PROJECT: WATER RESIDENCE 1. LAKESIDE
(name, address) HOLDEMAT BAY, WISCONSIN

State of: WISCONSIN
County of: MEDFLY

The undersigned, pursuant to Article 9 of the General Conditions of the Contract for Construction, AIA Document A201, hereby certifies that, except as listed below, he has paid in full or has otherwise satisfied all obligations for all materials and equipment furnished, for all work, labor, and services performed, and for all known indebtedness and claims against the Contractor for damages arising in any manner in connection with the performance of the Contract referenced above for which the Owner or his property might in any way be held responsible.

EXCEPTIONS: (If none, write "None". If required by the Owner, the Contractor shall furnish bond satisfactory to the Owner for each exception.)

NONE

SUPPORTING DOCUMENTS ATTACHED HERETO:

1. Consent of Surety to Final Payment. Whenever Surety is involved, Consent of Surety is required. AIA DOCUMENT G707, CONSENT OF SURETY, may be used for this purpose.
   Indicate attachment: (yes ☒) (no).

The following supporting documents should be attached hereto if required by the Owner:

1. Contractor's Release or Waiver of Liens, conditional upon receipt of final payment.

2. Separate Releases or Waivers of Liens from Subcontractors and material and equipment suppliers, to the extent required by the Owner, accompanied by a list thereof.

3. Contractor's Affidavit of Release of Liens (AIA DOCUMENT G706A).

CONTRACTOR: PHILIP PA. TRENELTYN INC.

Address: WITTS END, WISCONSIN

BY: *Pda Trenelyn*

Subscribed and sworn to before me this

24TH day of NOVEMBER 1981

Notary Public: *C. R. Name*

My Commission Expires: JUNE 1982

ONE PAGE

CONTRACTOR'S AFFIDAVIT OF RELEASE OF LIENS

AIA DOCUMENT G706A

OWNER ☑
ARCHITECT ☑
CONTRACTOR ☐
OTHER ☐

ARCHITECT'S PROJECT NO: 203

CONTRACT FOR: SINGLE FAMILY DWELLING

CONTRACT DATE: 15 MAY 1981

TO (Owner)

ELLEN I. WATER
P.O. BOX 514
HOLDEMAT BAY, WISCONSIN

PROJECT: WATER RESIDENCE 1, LAKESIDE
(name, address) HOLDEMAT BAY, WISCONSIN

State of: WISCONSIN
County of: MEDFLY

The undersigned, pursuant to Article 9 of the General Conditions of the Contract for Construction, AIA Document A201, hereby certifies that to the best of his knowledge, information and belief, except as listed below, the Releases or Waivers of Lien attached hereto include the Contractor, all Subcontractors, all suppliers of materials and equipment, and all performers of Work, labor or services who have or may have liens against any property of the Owner arising in any manner out of the performance of the Contract referenced above.

EXCEPTIONS: (If none, write "None". If required by the Owner, the Contractor shall furnish bond satisfactory to the Owner for each exception.)

NONE

CONTRACTOR: PHILIP DA TRENCHYN INC.

Address: WITTS END, WISCONSIN

BY: _(signature)_

Subscribed and sworn to before me this
24TH day of NOVEMBER 1981

Notary Public: _(signature)_

My Commission Expires: JUNE 1982

SUPPORTING DOCUMENTS ATTACHED HERETO:

1. Contractor's Release or Waiver of Liens, conditional upon receipt of final payment.

2. Separate Releases or Waivers of Liens from Subcontractors and material and equipment suppliers, to the extent required by the Owner, accompanied by a list thereof.

Consent of Surety Company to Final Payment

CONSENT OF SURETY COMPANY TO FINAL PAYMENT

AIA DOCUMENT G707

OWNER ☒
ARCHITECT ☒
CONTRACTOR ☒
SURETY ☐
OTHER

PROJECT: WATER RESIDENCE J. LAKESIDE
(name, address) HOLDEMAT BAY, WISCONSIN

TO (Owner)

⌈ ELLEN I. WATER
P.O. BOX 314
⌊ HOLDEMAT BAY, WISCONSIN

CONTRACTOR: PHILIP DA TRENCHTN INC.
WITTS END, WISCONSIN

ARCHITECT'S PROJECT NO: 203

CONTRACT FOR: SINGLE FAMILY DWELLING

CONTRACT DATE: 15 MAY 1981

In accordance with the provisions of the Contract between the Owner and the Contractor as indicated above, the

(here insert name and address of Surety Company)

ACME GUARANTEE INC.
MILWAUKEE, WISCONSIN   , SURETY COMPANY,

on bond of (here insert name and address of Contractor)

PHILIP DA TRENCHTN INC.
WITTS END, WISCONSIN   , CONTRACTOR,

hereby approves of the final payment to the Contractor, and agrees that final payment to the Contractor shall not relieve the Surety Company of any of its obligations to (here insert name and address of Owner)

ELLEN I. WATER   , OWNER,
P.O. BOX 314
HOLDEMAT BAY, WISCONSIN

as set forth in the said Surety Company's bond.

IN WITNESS WHEREOF,
the Surety Company has hereunto set its hand this   EIGHTEENTH   day of DECEMBER   1981

ACME GUARANTEE INC.
Surety Company

_____
Signature of Authorized Representative

PRESIDENT
Title

Attest: H. Douglas.
(Seal):

NOTE: This form is to be used as a companion document to AIA DOCUMENT G706, CONTRACTOR'S AFFIDAVIT OF PAYMENT OF DEBTS AND CLAIMS, Current Edition

AIA DOCUMENT G707 • CONSENT OF SURETY COMPANY TO FINAL PAYMENT • APRIL 1970 EDITION • AIA®
© 1970 • THE AMERICAN INSTITUTE OF ARCHITECTS, 1735 NEW YORK AVE., NW, WASHINGTON, D.C. 20006   ONE PAGE

Certificate of Substantial Completion

CERTIFICATE OF SUBSTANTIAL COMPLETION

AIA DOCUMENT G704

Distribution to:
- OWNER ☒
- ARCHITECT ☒
- CONTRACTOR ☒
- FIELD ☐
- OTHER ☐

PROJECT: (name, address) WATER RESIDENCE J. LAKESIDE HOLDEMAT BAY, WISCONSIN

TO (Owner):
ELLEN I. WATER
P.O. BOX 314
HOLDEMAT BAY, WISCONSIN

DATE OF ISSUANCE: 28TH NOV. 1981

PROJECT OR DESIGNATED PORTION SHALL INCLUDE:

WATER RESIDENCE INCLUDING ALL EXTERIOR LANDSCAPING

ARCHITECT: FAIR & SQUARE A.I.A.

ARCHITECT'S PROJECT NUMBER: 203

CONTRACTOR: PHILIP DA TRENCHIN

CONTRACT FOR: SINGLE FAMILY DWELLING

CONTRACT DATE: 15 MAY 1981

The Work performed under this Contract has been reviewed and found to be substantially complete. The Date of Substantial Completion of the Project or portion thereof designated above is hereby established as

which is also the date of commencement of applicable warranties required by the Contract Documents, except as stated below.

DEFINITION OF DATE OF SUBSTANTIAL COMPLETION

The Date of Substantial Completion of the Work or designated portion thereof is the Date certified by the Architect when construction is sufficiently complete, in accordance with the Contract Documents, so the Owner can occupy or utilize the Work or designated portion thereof for the use for which it is intended, as expressed in the Contract Documents.

A list of items to be completed or corrected, prepared by the Contractor and verified and amended by the Architect, is attached hereto. The failure to include any items on such list does not alter the responsibility of the Contractor to complete all Work in accordance with the Contract Documents. The date of commencement of warranties for items on the attached list will be the date of final payment unless otherwise agreed to in writing.

FAIR & SQUARE A.I.A.
ARCHITECT   BY B. Fair.   DATE 28TH NOV. 81

The Contractor will complete or correct the Work on the list of items attached hereto within _____ days from the above Date of Substantial Completion.

PHILIP DA TRENCHIN
CONTRACTOR   BY P. da Trenchin   DATE 1ST DEC. 81

The Owner accepts the Work or designated portion thereof as substantially complete and will assume full possession thereof at _____ (time) on _____ (date).

ELLEN I. WATER
OWNER   BY E. I. Water   DATE 7TH DEC. 81

The responsibilities of the Owner and the Contractor for security, maintenance, heat, utilities, damage to the Work and insurance shall be as follows:

(Note—Owner's and Contractor's legal and insurance counsel should determine and review insurance requirements and coverage; Contractor shall secure consent of surety company, if any.)

SECURITY, MAINTENANCE, HEAT, UTILITIES AND INSURANCE SHALL BECOME THE RESPONSIBILITY OF THE OWNER AFTER 28TH NOV. 81.

Invoice for Architectural Services

INVOICE FOR ARCHITECTURAL SERVICES

DATE: DECEMBER 17, 1981

INVOICE NO: 10

ARCHITECT'S
PROJECT NO: 205

PROJECT WATER RESIDENCE 1 LAKESIDE
(Name, address) HOLDEMAT BAY, WISCONSIN

TO: ELLEN I. WATER
P.O. BOX 314
HOLDEMAT BAY, WISCONSIN

In accordance with the Owner-Architect Agreement dated 15 MARCH 1981
there is due at this time for architectural services and reimbursable items on the above Project, for the period ending
DECEMBER 15, 1981 the sum of 1,350.00
ONE THOUSAND THREE HUNDRED AND FIFTY Dollars ($1,350.00)

The above amount shall become due and payable 14 days from the date hereof.

INTEREST ON OVERDUE ACCOUNTS SHALL ACCRUE AT TEN PERCENT (10 %) PER WEEK

The present status of the account is as follows:

SCHEMATIC DESIGN PHASE @ 15% = $1,350.00 @ 100% COMPLETED AND PAID

DESIGN DEVELOPMENT PHASE @ 20% = $1,800.00 @ 100% COMPLETED AND PAID

CONSTRUCTION DOCUMENT PHASE @ 40% = $3,600.00 @ 100% COMPLETED AND PAID

BIDDING PHASE @ 5% = $450.00 @ 100% COMPLETED AND PAID

CONSTRUCTION PHASE @ 20% = $1,800.00 @ 100% COMPLETED AND PAID

$1,800.00 × 75% = $1,350.00
NO PREVIOUS BALANCE

ARCHITECT: FAIR & SQUARE A.I.A.

BY: B. Fair.

ADDRESS: HOLDEMAT BAY, WISCONSIN

AIA FMS THE AMERICAN INSTITUTE OF ARCHITECTS
FINANCIAL MANAGEMENT SYSTEM

AIA® FORM F5002 · INVOICE FOR ARCHITECTURAL SERVICES
THE AMERICAN INSTITUTE OF ARCHITECTS, 1735 NEW YORK AVENUE, N.W., WASHINGTON, D.C. 20006

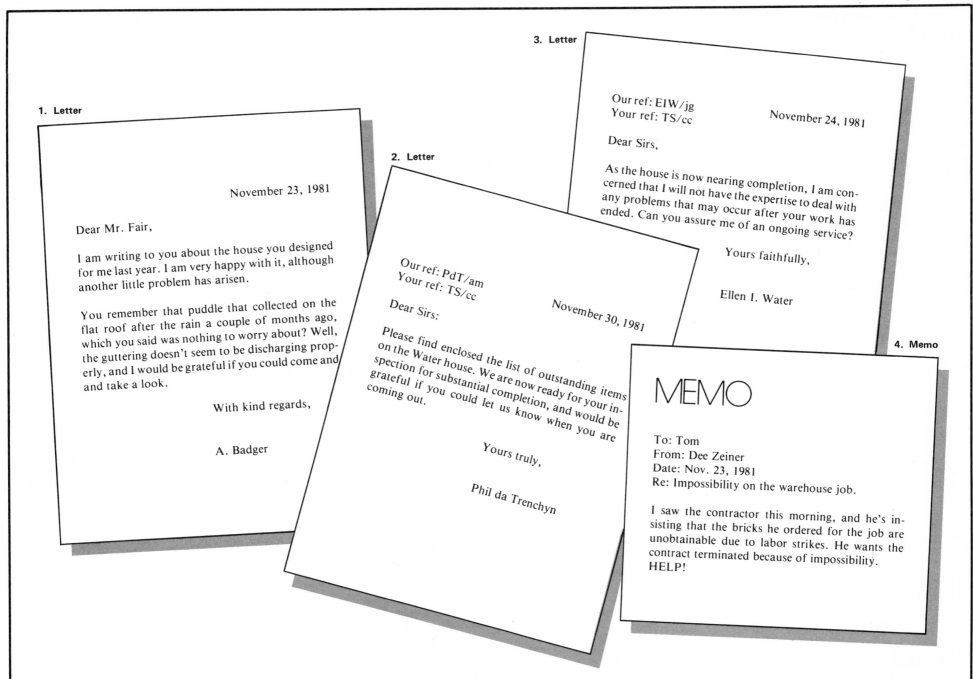

1. Letter

November 23, 1981

Dear Mr. Fair,

I am writing to you about the house you designed for me last year. I am very happy with it, although another little problem has arisen.

You remember that puddle that collected on the flat roof after the rain a couple of months ago, which you said was nothing to worry about? Well, the guttering doesn't seem to be discharging properly, and I would be grateful if you could come and and take a look.

With kind regards,

A. Badger

2. Letter

Our ref: PdT/am
Your ref: TS/cc

November 30, 1981

Dear Sirs:

Please find enclosed the list of outstanding items on the Water house. We are now ready for your inspection for substantial completion, and would be grateful if you could let us know when you are coming out.

Yours truly,

Phil da Trenchyn

3. Letter

Our ref: EIW/jg
Your ref: TS/cc

November 24, 1981

Dear Sirs,

As the house is now nearing completion, I am concerned that I will not have the expertise to deal with any problems that may occur after your work has ended. Can you assure me of an ongoing service?

Yours faithfully,

Ellen I. Water

4. Memo

MEMO

To: Tom
From: Dee Zeiner
Date: Nov. 23, 1981
Re: Impossibility on the warehouse job.

I saw the contractor this morning, and he's insisting that the bricks he ordered for the job are unobtainable due to labor strikes. He wants the contract terminated because of impossibility. HELP!

Action Taken

1. Letter and diary insert

November 26, 1981

Our ref: BF/cc

Dear Mr. Badger:

Thank you for your letter of November 23. I am sorry to hear you still feel that you have problems with your roof. As you will recall on the three previous occasions I have visited your house, I found nothing to worry about—perhaps a simple maintenance job is all that is required.

However, we will be happy to consult with you again on the matter if you wish, and enclose a copy of our scale of payments together with details of fees and expenses.

We look forward to hearing from you,

Yours sincerely,

Fair and Square

Letter

Diary insert

Date: 11/26/81
Re: Badger's continuing requests for advice.

This will be the fourth trip in the last few months and nothing is really wrong with the house—it was all cleared up under the contractor's warranty period. Maybe now we have to ease off the good will and politely but firmly put the relationship back onto a financial footing.

2. Diary insert

Date: Nov. 27, 1981
Re: Water House: Substantial Completion

Inspected the Water house yesterday for substantial completion. Made a few additions to the punch list, work on which is to be completed by January 10, 1982.

Forms for substantial completion ready to go out.

4. Memo

MEMO

To: Dee
From: Tom
Date: 11/25/81
Re: Impossibility

Unobtainable (?) bricks. I'm sure other alternatives must be available (other suppliers, change of specifications, etc.) before we have to resort to termination.

I'll go out to the site and check on this and then keep an eye on things. The contractor seems very keen to end the contract over nothing much, and there *was* that business with not paying the subcontractors and suppliers.

3. Letter

Our ref: TS/cc
Your ref: EIW/jg

November 26, 1981

Dear Ms. Water:

Thank you for your letter of November 24. As you know, any minor problems that may show up during the first 12 months after substantial completion will be taken care of by the contractor under the warranty expressed in the building contract.

If you like, we could provide inspection services of the work both before and after the contractor has completed any extra work to ensure its adequate completion. It may also be in your interests for us to make a thorough inspection of the house just before the warranty period expires to prepare a list of any work that should be completed for conformance with the contract documents. We would also then check the quality of this work upon completion. Our charges for additional services for such work are outlined in our agreement.

Please let us know your decision in the matter when the time comes. In the mean time, we have enjoyed working with you on this project and hope that you and your family enjoy your new home.

Yours sincerely,

Fair and Square

SECTION 8.
ARBITRATION

Contents   **Page:**

For a number of reasons, not all building contracts are fully performed as intended. A contract can be terminated in a variety of ways, e.g., by agreement (see page 73), but the AIA General Conditions make special provision for the unilateral termination of the contract by either the owner or the contractor in the event of specified circumstances.

Termination by the Owner (AIA Document A201, Article 14.2.1)

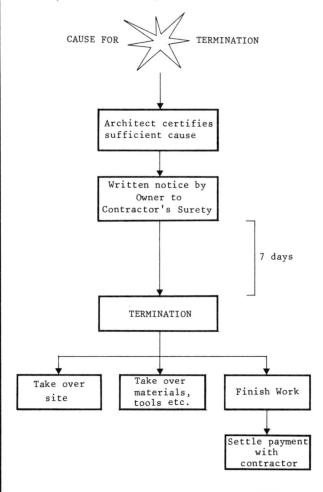

The owner may be permitted to terminate the contract:

- If the contractor is adjudged bankrupt
- If the contractor makes a general assignment for the benefit of his/her creditors
- If a receiver is appointed on account of the insolvency
- If the contractor persistently or repeatedly fails to supply properly skilled workmen or proper materials (unless an extension has been granted)
- If the contractor fails to make prompt payment to subcontractors and/or suppliers
- If the contractor persistently disregards laws, rules, ordinances, regulations, or orders of a public authority
- If the contractor is guilty of a substantial violation of one of the provisions of the contract documents.

Procedure

Under the AIA General Conditions, the owner must seek from the architect certification that sufficient cause exists to justify termination of the contract. It should be noted that the 1979 Federal Bankruptcy Act provides that trustees in bankruptcy may continue performance of contracts to which the bankrupt is a party. As a result of this change in the law, the AIA has advised that Article 14.2.1 has been effectively invalidated insofar as it relates to bankruptcy. Termination by the owner, therefore, should only be undertaken with the assistance of legal counsel, who should carefully review the circumstances in the light of the contract itself, and relevant state and federal law.

If the decision is made to terminate, seven days after written notice has been sent to the contractor and any surety, the owner may:

- Terminate the contract with the contractor
- Take possession of the site
- Take possession of all materials, equipment, tools, construction equipment, and machinery on the site owned by the contractor
- Finish the work in the most expedient way.

The contractor will not be entitled to any further payment of outstanding fees until the project is complete.

Costs

If the moneys owing to the contractor exceed the cost of finishing the work (including additional fees of the architect), the contractor will be reimbursed the difference. However, if the cost of finishing is higher than the amount owed to the contractor, the contractor will be liable to pay the excess, the sum of which must be certified by the architect.

Termination by the Contractor (AIA Document A201, Article 14.1.1)

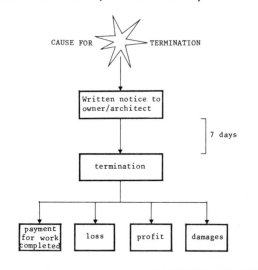

Termination

The contractor may also have the right to terminate the contract under the following circumstances:

If the work is stopped for a period of 30 days due to:

- Order of the Courts
- Public authority intervention
- The result of an act of government (e.g., declaration of national emergency, making materials unavailable)
- Through no act of the contractor or his/her subcontractors or agents

Or, if the work is stopped for 30 days:

- Because the architect has not issued a certificate for payment
- Because the owner had not paid the amount certified.

Procedure

After seven days written notice to the owner and the architect, the contractor may terminate the contract and recover from the owner:

- Payment for all work executed to date
- Proven loss sustained in the expenditure for materials, equipment, tools, construction equipment, and machinery
- Reasonable profit
- Damages.

Termination is a drastic step to take in the event of contractual disputes, and should be given extremely careful consideration. The aggrieved party should ensure that all procedures required by the contract documents and by relevant laws are strictly adhered to in order to prevent successful counterclaims.

Reference

Sweet, pp. 542-562.

If all other contractual mechanisms fail to provide satisfactory resolution of a dispute between the contracting parties, the introduction of a third party may be necessary to settle the matter.

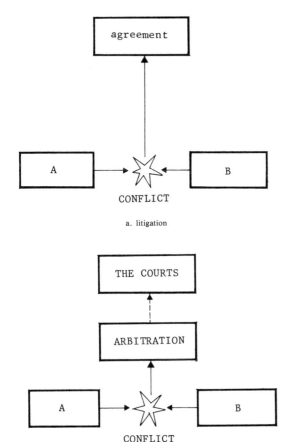

a. litigation

b. arbitration

The third party could be a civil-court judge if the normal court procedures are followed (see page 5). Alternatively, a dispute could be settled by the relatively less formal procedure of arbitration.

Whereas the courts form part of the United States' judicial system and are, therefore, subject to all of its procedural and administrative rules, disputes submitted to arbitration can be settled by an informal private hearing in the presence of whomever the parties choose. If agreement is not possible, a designated third party may select the arbitrator.

The arbitrator is usually someone with specific knowledge and experience in the field in which the dispute has arisen. In the building sector, this might be an architect, engineer, or other professional person who is usually a member of the American Arbitration Association, a national organization which operates a commercial panel that serves, among other things, the building industry.

Advantages of Arbitration

The major advantages of submitting a dispute to arbitration are:

a. Privacy
b. Convenience
c. Speed
d. Expense
e. Informality
f. Expertise.

a. Privacy

Trade secrets and reputations may be shielded from the public in a private arbitration. The courts, however, are public forums and privacy is generally not possible.

b. Convenience

Arbitration hearings can be held anywhere to suit the parties.

c. Speed

Disputes can be handled quickly, without the inconvenience of having to fit into a court's schedule. In projects where time is of the utmost importance, this can be a decisive factor.

d. Expense

Money might be saved in two ways:

1. The potentially lower cost of the hearing
2. The speedy resolution of the dispute.

e. Informality

Courtroom procedures may be dispensed with or modified at the discretion of the arbitrator.

f. Expertise

Difficult construction-oriented problems may be more readily understood by an arbitrator experienced in the construction field than by a professional judge.

Disadvantages of Arbitration

The major disadvantages of arbitration are:

a. Cost
b. Lack of legal expertise
c. No binding precedent.

a. Cost

Aside from the expense of legal counsel, expert witnesses, etc., the arbitrators fees must be paid together with the cost of hiring the place of the hearing. In the court system, the services of the judge and the use of the courtroom are not additional expenses.

Arbitration

b. Lack of Legal Expertise

Though knowledgeable in the field of the dispute, the arbitrator may be less well-informed in respect of the law than a professional judge.

c. No Binding Precedent

Each case submitted to arbitration is decided upon its own merits, without necessarily any regard to previous cases. This can make it difficult for the parties to ascertain the strength of their arguments.

When to Arbitrate

Parties may go to arbitration:

- By agreement after the dispute has arisen
- By agreement before the dispute arises (i.e., as a condition of the contract)
- By order of court (many states will enforce an agreement to arbitrate).

Agreement to arbitrate prior to a dispute occurring is the preferable method, and most building contracts provide for arbitration proceedings by stating that the parties agree to be bound by the decision of an arbitrator in the event of disagreement (AIA Document A201, Article 7.9.1). Many AIA standard forms of contract provide for arbitration, including the owner–architect agreements.

In addition to the agreement to arbitrate, both parties to AIA construction contracts agree to abide by the Construction Industry Arbitration Rules which are published by the American Arbitration Association. However, some state laws regarding arbitration vary, and this should be taken into account at the contract formation stage in case any modifications may be necessary to match state requirements. The assistance of legal counsel is advisable.

References

Sweet, pp. 563–593.
Walker, pp. 180–188.
Acret, pp. 316–334.

Either party to the AIA construction contract (not necessarily with the consent of the other party at the time of the dispute) may initiate arbitration proceedings by writing to the other party, with a copy to the architect, within the time allowed by the contract documents (AIA Document A201, Article 2.2.12). The letter usually includes:

- The reason for the dispute
- The amount involved
- The remedy sought

and represents Notice of Demand for Arbitration under the terms of the AIA contract. Two copies of the notice should be filed with the American Arbitration Association at their nearest regional office, together with the requisite filing fee. The other party may file an answer in duplicate with the American Arbitration Association (AAA) within 7 days of the notice, sending a copy of the answer to the originator of the proceedings. If the respondent chooses not to reply, there is nonetheless an assumption that the claim is denied.

Although arbitration proceedings are in progress, both parties are constrained by the contract to meet all their contractual obligations unless otherwise agreed in writing, or unless the reason for the arbitration is the breakdown of the contract itself.

Selection of the Arbitrator

An arbitrator may be selected:

- By agreement of the parties before the dispute
- By agreement of the parties during the dispute
- By reference to the American Arbitration Association.

In the latter case, the AAA sends a list of possible arbitrators to both parties who are given 7 days to delete any names they consider to be unacceptable, and list the remaining names in order of preference. The AAA then approaches an arbitrator (or arbitrators: a panel of three is often selected) on the basis of the amended lists. In the event that none of the names are acceptable, the AAA will appoint an arbitrator without submitting new lists.

Prior to accepting the appointment, the prospective arbitrator should assess his/her suitability for the case, and disclose all potential conflicts of interest (e.g., personal knowledge of one of the parties, or a financial interest in the dispute).

Pre-Hearing Procedures

A pre-hearing conference may be arranged at the parties' request, or if the AAA believes such a conference would be useful. The pre-hearing conference allows for an exchange of information, the stipulation of uncontested facts, and the agreement of administrative details such as:

- Locale
  This may be mutually agreed upon, but in the event of disagreement between the parties, the AAA will make a binding decision.
- Use of legal counsel
  This is acceptable in many states, but if one party decides to engage a legal representative, the other pary and the AAA must be notified at least three days prior to the hearing.
- Stenographic record
  If one of the parties requests a record of the proceedings, that party must bear the costs unless both parties agree to share the expense.

- Time and place
  The arbitrator decides the time and place of the hearing, and the AAA will notify the parties at least five days in advance.

The Hearing

The hearing should only be held when all the requisite documents have been exchanged. In the event of the refusal of one party to participate, or if there is an attempt to deliberately obstruct the proceedings, the arbitration may continue *ex parte* (i.e., on the proof of one party only) provided that the absent party has been notified in writing of his/her right to attend. An award may not be made simply on the basis of the absent party's default, and all relevant evidence should be heard by the arbitrator prior to making the award.

The hearing generally comprises the following stages:

- The oath of the arbitrator (if required)
- Recording of time, place, date of hearing, the parties present and statements of claim and response
- The arbitrator may ask for statements from both sides outlining the issues involved in the dispute
- The claimant will then present the claim, supported by proof in the form of testimony, exhibits, etc.
- The claimant's witnesses will be examined, cross-examined (by the respondent or counsel) and then re-examined by the claimant
- The respondent must then follow the same procedure for the defense and counterclaim (if any).

Arbitration Procedure

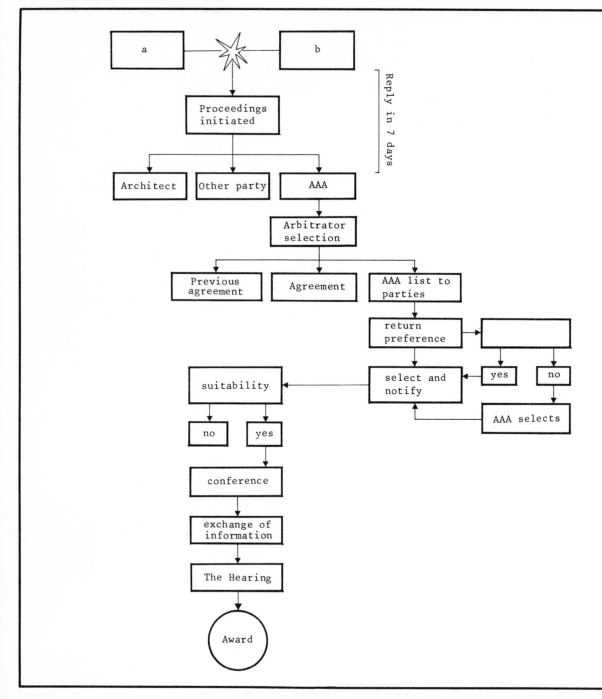

Inspection of property may be required, and both parties are generally given the opportunity to accompany the arbitrator. If no other proof is required or forthcoming, the arbitrator will close the proceedings and make a decision within the specified time (usually not later than 30 days after the hearing). No communication between the arbitrator and the parties to the dispute should take place, except through the AAA.

The Award

The award should be made in accordance with the relevant state law and will be sent to both parties simultaneously by the AAA. In some instances, the parties may be asked to deposit a sum with the AAA at the beginning of the arbitration proceedings to ensure payment of the arbitrator.

In the event that one of the parties refuses to accept the arbitrator's decision, application may be made to the courts to enforce the award.

Although arbitrations are carried out largely independently of the court system, the courts may have statutory power to reject or vacate the arbitrator's award:

- If the arbitrator exceeds his/her authority
- If there is evidence of corruption, fraud, or partiality
- If the arbitrator refuses to hear evidence of either party
- If the arbitration agreement is improper.

Modifications to the award may be allowed by the arbitrator if a party considers that a mistake has been made. However, awards are usually reaffirmed.

The Architect as Arbitrator

As noted earlier, the architect has a quasi-arbitral role in the administration of the construction contract. In addition, the architect's professional qualification and experience in the construction field implies a knowledge and expertise which might provide the basis for arbitration work. The American Arbitration Association has regional offices throughout the United States which may be contacted by architects wishing to apply for inclusion on the commercial panel.

The Architect as Expert Witness

It is possible that an architect may be called as an expert witness at an arbitration to give professional opinions regarding a building dispute. The expert witness is not usually personally involved in the dispute, and is paid for objective expertise and opinion which may be given in a written report or by oral testimony.

The Award

American Arbitration Association, Administrator
Commercial Arbitration Tribunal

Wesmey-Shovelgon Construction Co.
and
Acme Estates Inc.

Case Number 4136–1321–88

AWARD OF ARBITRATOR

I, THE UNDERSIGNED ARBITRATOR, having been designated in accordance with the Arbitration Agreement entered into by the above-named parties, and dated May 14, 1980, and having been duly sworn and having heard the proofs and allegations of the parties, AWARD as follows:

1. Within fourteen (14) days from the date of transmittal of this Award to the parties, Acme Estates Inc. shall pay to Wesmey-Shovelgon Construction Co. the sum of FOURTEEN THOUSAND THREE HUNDRED AND FORTY-SIX DOLLARS ($14,346.00) plus interest thereon at the rate of eleven and one quarter per cent (11.25 per cent) per annum from the date when construction work was stopped by Acme Estates Inc., that being May 15, 1981, until August 21, 1981.

2. The counterclaim of Acme Estates Inc. against Wesmey-Shovelgon Construction Co. is hereby denied.

3. The administrative fees of the American Arbitration Association amounting to SEVEN HUNDRED AND FIFTY-FOUR DOLLARS AND THIRTY FIVE CENTS ($754.35) shall be borne entirely by Acme Estates Inc.

4. This award is in full and final settlement of all claims and counterclaims submitted to the arbitration.

signed: _____

Arbitrator _____

Date: _____

Notarized: _____

Note: The execution of the award may vary according to the legal requirements of the state in which the arbitration takes place.

The American Institute of Architects
1735 New York Avenue, N.W., Washington, DC
20006

American Arbitration Association
140 W. 51st St., New York, NY 10020

American Institute of Interior Designers
730 5th Ave., New York, NY 10019

American Institute of Landscape Architects
501 E. San Juan Ave., Phoenix, AZ 85012

American Insurance Association
85 John St., New York, NY 10038

American National Standards Institute Inc.
1430 Broadway, New York, NY 10018

American Society for Testing and Materials
1916 Race St., Philadelphia, PA 19103

American Society of Civil Engineers
345 E. 47th St., New York, NY 10017

American Society of Mechanical Engineers Inc.
345 E. 47th St., New York, NY 10017

Associated Builders and Contractors Inc.
PO Box 698, Glen Burnie, MD 21061

Associated General Contractors of America
1957 East St., N.W., Washington, DC 20006

Building and Code Administrators International
Inc.
1313 E. 60th St., Chicago, IL 60637

Building Research Institute
2101 Constitution Ave., N.W., Washington, DC
20418

Copyright Office
Library of Congress, Washington, DC 20559

International Conference of Building Officials
50 S. Los Robles, Pasadena, CA 91101

Interprofessional Commission on Environmental
Design
917 15th St., N.W., Washington, DC 20005

Model Codes Standardization Council
303 W. Commonwealth Ave., Fullerton, CA
92632

National Association of Building Manufacturers
1701 18th St. N.W., Washington, DC 20009

National Building Products Association
120-44 Queen's Blvd., Kew Gardens, NY 11415

National Housing Producers Association
900 Peachtree St., N.E., Atlanta, GA 30309

National Planning Association
1606 New Hampshire Ave., N.W., Washington,
DC 20009

National Society of Professional Engineers
2029 K St., N.W., Washington, DC 20006

The following books have been used as reference sources throughout the text and are recommended:
Legal Aspects of Architecture, Engineering and the Construction Process. Sweet, J., St. Paul: West Publishing Co. 2nd Edition. 1977.
Legal Pitfalls in Architecture, Engineering and Building Construction. Walker, N., Walker, E.N., Rohdenberg, K. New York: McGraw-Hill Books Co. 2nd Edition. 1979.
Architects and Engineers: Their Professional Responsibilities. Acret, J. Colorado Springs: Shepard's Inc. and New York: McGraw-Hill Book Co. 1977.
The Architect's Handbook of Professional Practice. The American Institute of Architects.

Further Reading

These books are recommended for further reference in the field of legal and professional matters related to architectural practice:
AIA Building Construction Legal Citator. Spencer, Whalen and Graham. Washington, D.C.: The AIA. Revised Edition. 1980.
Architectural Engineering and Law. Tomson, B., Coplan, N. New York: Reinhold Publishing Corp. 2nd Edition. 1967.
It's the Law! Recognizing and Handling the Legal Problems of Public and Private Construction. Tomson, B. Great Neck, N.Y.: Channel Press. 1960.
A Building Code Primer. Correale, W. H. New York: McGraw-Hill Book Co. 1979.
Construction Contracts. Collier, K. Reston, Va.: Reston Publishing Co., Inc. 1979.
Preventing and Solving Construction Contract Disputes. Hohns, H.M. New York: Van Nostrand Reinhold Co. 1979.
Construction Delay. O'Brien, J.J. Boston Mass.: Cahner's Books International, Inc. 1976.

Building Contracts for Design and Construction. Hauf, H.D. New York: John Wiley and Sons. 2nd Edition. 1976.
Professional Practice in Architecture. Orr, F. New York: Van Nostrand Reinhold, 1982.
Construction Claims. Rubin, R., Guy, S., Maevis A., Fairweather, V. New York: Van Nostrand Reinhold, 1982.

For current information regarding law and practice, reference may be made each month to *Architectural Record* ("Legal Perspectives") and *Progressive Architecture* ("It's the Law").

The following AIA forms have been filled out and located in the text:

ab initio	From the beginning
bona fide	In good faith
caveat emptor	Let the buyer beware
ejusdem generis	Of the same type
estoppel	a rule of evidence which prevents a person from denying or asserting a fact owing to a previous act
ex parte	Upon the application of
ignorantia juris non excusat	Ignorance of the law is no excuse
in personam	Against a person, i.e., not against everyone
in rem	Against a thing, i.e., applicable to everyone
inter se	Among themselves
obiter dicta	Things said by the way
per se	By itself
prima facie	On first view
quantum meruit	As much as he deserves
ratio decidendi	Reason for the decision
res ipsa loquitur	The thing speaks for itself
stare decisis	To stand by past decisions
sui juris	Of legal capacity
tortfeasor	One liable for a civil wrong, except re: contract or trust
uberrimae fidei	Of the utmost good faith
ultra vires	Beyond one's powers
volenti non fit injuria	No wrong can be done to one who consents to the action

List of Diagrams

Index

Index

Index